CLIMATE CHANGE AND ITS CAUSES, EFFECTS AND PREDICTION

FLOODING AND CLIMATE CHANGE

SECTORIAL IMPACTS AND ADAPTATION STRATEGIES FOR THE CARIBBEAN REGION

CLIMATE CHANGE AND ITS CAUSES, EFFECTS AND PREDICTION

Additional books in this series can be found on Nova's website under the Series tab.

Additional e-books in this series can be found on Nova's website under the e-book tab.

CLIMATE CHANGE AND ITS CAUSES, EFFECTS AND PREDICTION

FLOODING AND CLIMATE CHANGE

SECTORIAL IMPACTS AND ADAPTATION STRATEGIES FOR THE CARIBBEAN REGION

DAVE D. CHADEE
JOAN M. SUTHERLAND
AND
JOHN B. AGARD
EDITORS

nova publishers
New York

For permission to use material from this book please contact us:
Telephone 631-231-7269; Fax 631-231-8175
Web Site: http://www.novapublishers.com

Library of Congress Cataloging-in-Publication Data

ISBN: 978-1-62948-317-7

Published by Nova Science Publishers, Inc. † New York

CONTENTS

PREFACE

At the beginning of the 21st century the science of climate change reached a pivotal stage during which the evidence of its reality had become overwhelming and both developed and developing countries similarly experienced the impacts of extreme global climate-change events such as flooding, heat waves and drought. However, what was instructive was the rate of recovery of developed versus developing countries and their contrasting rates of implementing adaptation measures to reduce future impacts. In poor developing countries like Small Islands Developing States (SIDS), recovery periods were notably longer as a result of limited financial resources, lack of political will, need for regional and international aid, the lack of technical expertise for reconstruction of homes and villages, and the failure to develop adaptation measures which could potentially prevent or reduce the impact of future extreme events. This failure may be partially explained by the fact that economic losses from natural disasters in SIDS (as a percentage of GDP) have been >1% and 8% in the worst cases, as opposed to <0.1% in high-income countries (IPCC, 2012).The Caribbean islands represent a prime example of developing countries with inadequate resilience to cope with extreme events. Most islands' coastlines are vulnerable to sea-level rise (Pulwarty et al. 2010, Nurse 2011), and, since most villages and townships, and much of the agriculture sector, are located along the coast, they are, therefore exposed to sea-level rise and storm surges as well as the ravages of hurricanes (IPCC 2013). However, regional institutions have held consultations with major stakeholders and have tried to identify and address many of these obvious vulnerabilities, but their success in building resilience is, thus far limited. Some gray literature exists within the Caribbean region but this data is often unreliable or lacking the scientific rigor required for data that can be used to inform policy and ultimately planning.

Within the Caribbean region three books have been written relating to global climate change and the vulnerability of SIDS. The first examined the available literature on vulnerability in the Caribbean in the context of both 'globalisation' and climate change (McGregor et al. 2009), while the edited book by Baban (2008) addressed the hydrology and flow dynamics of rivers which cause flooding but did not address the issue of extreme events and how adaptation strategies should be approached to build resilience. The third book addressed the key issues of dengue and climate change in the Caribbean region and provided the evidence required for developing early warning systems for reducing both the vector, *Aedes aegypti* mosquitoes and transmission of the dengue virus (Chen et al. 2006) but little follow-up action has been undertaken to develop the adaptation measures recommended.

The current book has built on the work described in these previous publications and has consciously adopted a sectorial approach with the purpose of to providing sufficient within-sector information for actors in other sectors to appreciate how their planned or proposed adaptation actions can have wider benefits. In other words 'starting again' in the iterative process between various stakeholders and their differing approaches that is necessary to reconcile these, and to ensure the development of cost-effective "low regrets"[1] adaptation strategies to counter climate change vulnerabilities. The book begins by addressing the key issues of vulnerability relating to global and projected climate change in the Caribbean (Chapters 1 and 2). Important vulnerabilities are described in agriculture (chapter 3) and in the wetland and mangrove ecosystems described in chapters 4 and 5. Chapters 6 & 7 provide recent relevant scientific research data on health and social attitudes to flooding vulnerability. Hard and soft engineering solutions to flooding management are dealt with in chapters 7 & 8, while in chapter 9 the adaptation solutions to Jamaica's serious flooding problem are described. The chapter on the law relating to flooding in Trinidad and Tobago is instructive in that it highlights the need for a cross-sectorial solution to man's exacerbations of flooding vulnerabilities. The final chapter provides an explanation of the scenarios used in the most recent IPCC report (IPCC, 2013), and how international science and modeling will allow future assessment of the cost effectiveness of adaptation and mitigation efforts.

This book is intended for both undergraduate and post graduate students pursuing climate change courses and programs at the University of the West Indies, Cave Hill, Mona and St. Augustine campuses, as well as for government technocrats and other relevant stakeholders. The various chapters are written by current academics from all three campuses, together with regional experts. This book, therefore, adds to a growing body of work by Caribbean scientists/experts tackling Caribbean issues and developing adaptation measures relevant, not only to the region, but to the rest of the world. The authors recognize that flooding affects many other sectors e.g., transportation and security. Therefore, as other rigorously collected and scientifically researched data become available, and new adaptation strategies are developed, it is envisaged that this information will be included in a second edition of this book thus making it as policy- relevant especially for SIDS as we believe this edition to be.

REFERENCES

Baban, S.M.J. (2008). Enduring Geohazards in the Caribbean: Moving from the reactive to the Proactive. The University of the West Indies Press, Trinidad.

Chen, A.A., Chadee, D.D. and Rawlins, S.C. (2006). Climate Change Impacts on Dengue: The Caribbean Experience. Phoenix Press, Jamaica.

IPCC (2012). Summary for Policymakers. In: Managing the Risks of Extreme Events and Disasters to Advance Climate Change Adaptations [Field, C.B., V. Barros, T.F. Stocker, D. Qin, D.J. Dokken, K.L. Ebi, M.D. Mastrandrea,K.J. Mach, G.-K. Plattner, S.K. Allen, M. Tignor, and P.M. Midgley (eds)] A Special Report of Working Groups I and II of the Intergovernmental Panel on Climate Change. Cambridge University Press, Cambridge, UK, and New York, NY, USA, pp. 1-19.

[1] Used in IPCC(2012) Special Report: Managing the Risks of Extreme events and disasters to Advance Climate Change Adaptation

IPCC (2013). The Physical Science Basis. IPCC Working Group 1 Contribution to AR5. Working Group 1 Contribution to the 5th Assessment of the International Panel on Climate Change, Cambridge University Press, Cambridge, UK.

McGregor, D, Dodman, D and Barker, D (2009). Global Change and Caribbean Vulnerability. University of the West Indies Press, Jamaica

Nurse, L.A (2011). The implications of global climate change for fisheries management in the Caribbean. Climate and Development, 3: 228-241.

Pulwartry, R.S., Nurse, L.A. and Trotz, U.O. (2010) Caribbean Islands in a Changing Climate. Environmental Science and Policy for Sustainable Development. Nov-Dec 2010 p. 9.

Editors:

Dave D. Chadee Ph.D., D.Sc.
Professor of Environmental Health
Department of Life Science
University of the West Indies, St Augustine
Trinidad, West Indies

Joan M. Sutherland, Ph.D.
Consultant/Ecophysiologist
Department of Life Science
University of the West Indies, St Augustine
Trinidad, West Indies

John B. Agard, Ph.D.
Professor of Tropical Island Ecology
Department of Life Science
University of the West Indies, St Augustine
Trinidad, West Indies

ACKNOWLEDGMENTS

The preparation of this book was aided by the 2013 Anthony N Sabga Caribbean Awards for Excellence Prize in Science & Technology which was awarded to Prof Dave Chadee. The St Augustine Campus, University of the West Indies also provided a study and travel grant (2013) which facilitated the completion of this book. The authors also thank Dr. Fiona Williams, Reader, School of Medicine, Division of Population Health Sciences, The University of Dundee, Dundee DD1 9SY, Scotland, UK for providing the space and atmosphere required for writing and editing some of the chapters of this book and we thank Dr. Judy Raymond, Editor in Chief of the Trinidad and Tobago Guardian for permission to republish photographs. We also thank our families for their support and encouragement during the preparation of this book.

In: Flooding and Climate Change
Editors: D. D. Chadee, J. M. Sutherland and J. B. Agard
ISBN: 978-1-62948-317-7

Chapter 1

INTRODUCTION TO FLOODING AND CLIMATE CHANGE IN THE CARIBBEAN

Dave D. Chadee* and Joan M. Sutherland
Department of Life Sciences, The University of the West Indies,
St. Augustine, Trinidad, West Indies

ABSTRACT

The factors associated with climate change are reviewed and discussed in the context of flooding and its impacts on Small Island Developing States (SIDS) within the Caribbean region. The natural factors which contribute to flooding including rainfall, hurricanes, storm surges, tidal flooding and tsunamis; and the anthropogenic factors such as deforestation, quarrying, urbanization, poor maintenance of canals, rivers and water courses, and the destruction of natural coastal barriers such as mangroves, are discussed with respect to the development of appropriate adaptation strategies. The vulnerabilities to climate change of the differing island types are also identified and reviewed.

1. INTRODUCTION

There is overwhelming evidence that the world's climate is changing (IPCC 2007, 2012, 2013) and the debate no longer centers on whether the primary triggers are anthropogenic, or are due to natural or cyclical changes in the earth's atmosphere, or to a combination of these factors. Earlier evidence showed that the earth's climate was affected by natural processes, including changes in volcanic activity, solar radiation output and the earth's orbit around the sun (IPCC 2012). Volcanic eruptions are often considered cyclical and can have short term effects on the climate, while solar irradiance has contributed to climate change over the last century. However, since the Industrial Revolution, the addition and increase of 'greenhouse gas es ' (water vapour, methane, carbon dioxide, nitrous oxide, CFCs), that trap solar radiation within the earth's atmosphere, has increased the scale of change to approximately 10

* E-mail: Chadee@tstt.net.tt; Dave.chadee@sta.uwi.edu.

times that of the Sun 's output. In 2013, IPCC provided the evidence that atmospheric CO_2 was the main driver of climate change since 1750 and that it is *"extremely likely*[1] that human influence has been the dominant cause of the observed warming since the mid-20th century ". In addition, there is evidence that the "Global surface temperature for the end of the 21st century will *likely*[2] exceed 1.5°C relative to 1850 to 1900 for all RCP (Representative Concentration Pathways) scenarios[3] except RCP2.6[4], and *more likely than not* to exceed 2°C for RPC 4 (see Chapter 12).

Therefore, we live in a world which cannot continue to adopt a "business as usual approach" as there is overwhelming evidence to support the fact that the observed changes have impacted various sectors including agriculture, ecosystems and their services, health, and tourism, and there is currently a suite of adaptation measures which can be applied by countries to 'slow-down' these impacts. However, while many of these adaptation measures may be effective under local conditions, they are not generic ('one-size -fits-all') tools, because of differences in geography and socio-economic conditions. The efficacy of these adaptation measures depends on differences and changes in adaptive capacity and/or the resilience of the affected individuals, communities and countries.

The specific impacts anticipated by climate change scientists are listed below.

- *Sea level rise*
  A rise in global sea level of 19 centimetres (7.5 inches) occurred between 1901 and 2010. Based on current acceleration levels, the projected rise for the current century will be almost double that observed during the last century (IPCC 2013).
- *Global temperature rise*
  Global models have shown that the earth's surface temperature has seen warming since 1880 but with significantly faster warming occurring since the 1970s. The warmest years have occurred since 1981 and, even though a solar output decline was reported in 2007-2009, surface temperatures continue to increase (Peterson et al. 2009, Allison et al., 2009).
- *Warming oceans*
  The top 700 metres (circa 2,300 feet) of the ocean have absorbed significant amounts of heat generated/trapped as a result of global warming and has shown a warming of 0.302 degrees Fahrenheit since 1969 (Allison et al., 2009).
- *Shrinking ice sheets*
  Reports from the NASA's Gravity Recovery and Climate Experiment show "Greenland lost 150 to 250 cubic kilometers (36 to 60 cubic miles) of ice per year between 2002 and 2006, while Antarctica lost about 152 cubic kilometers (36 cubic miles) of ice between 2002 and 2005"
- *Declining Arctic sea ice*
  The extent and thickness of the Arctic sea ice have both decreased over the last 50 years (Polyak et al., 2009).
- *Glacial retreat*

---

[1] 95-100% probability (IPCC, 2012).

[2] 66-100% probability (IPCC,2012).

[3] See chapter 12.

[4] Level of radiative forcing (see IPCC2013), in watts per square. The RCPs 2.6,4.5,6,and 8.5 are equivalent to CO_2 concentrations of 490,650,850 and >1370 ppm respectively.

Glaciers in the Alps, Himalayas, Andes, Rockies, Alaska and Africa are retreating at a rapid rate (IPCC 2007).

- *Extreme events*
  Globally the numbers of recorded days with high temperatures and heat waves have been increasing, while the number of low temperature events has been decreasing. Flooding episodes have also increased due to an increase in the number of intense rainfall events (IPCC 2007, IPCC 2012).
- *Ocean acidification*
  The acidity of the surface ocean water has been modified by anthropogenic factors, largely beginning with the industrial revolution, when a 30% increase in acidity occurred. The on-going emission of carbon dioxide into the atmosphere by human activity is continuing to be absorbed by the oceans. It is estimated that the amount of carbon dioxide absorbed by the upper layer of the ocean increases by 2 billion tons per year (Sabine et al., 2004, IPCC 2007, IPCC 2012).

In the following chapters we review the role of flooding which is an extreme event increasingly associated with climate change, and examine the potential impacts on Small Island Developing States (SIDS) within the Caribbean region.

2. Causes of Flooding

Flooding occurs when the volume of water within water courses or rivers exceeds its 'carrying capacity' and occurs along low-lying areas especially along coastlines, many of which are constantly inundated and have limited capacity for extra water absorption.

Flooding in the Caribbean region occurs:

- during and after meteorological events such as storms when intense and prolonged rainfall occurs;
- during coastal storm surges or during peak high tide;
- when peak high tide or seiche (a wave that oscillates in lakes, bays, or gulfs from a few minutes to a few hours as a result of seismic or atmospheric disturbances) coincide with episodes of heavy rainfall which drain into the same coastal margin;
- when natural storm barriers such as mangroves are destroyed by man and/or natural forces; and
- when seismic waves affect the ocean floor, triggering tsunamis.

Table 1.1. Summary of the main possible causes of flooding within the Caribbean region

Natural	Artificial (Man-made)
Rainfall	Deforestation
Hurricanes	Quarrying/ Blasting
Storm surges	Urbanization
Tidal flooding	Canals and Poor water course maintenance
Tsunamis (seismic waves)	Destruction of mangroves and other natural barriers along the coastal zone

The principal causes of flooding can be divided into two categories: *natural* and *man - made*. These are summarized in Table 1.

2.1. Natural Causes of Flooding

2.1.1. Rainfall

The observed changes in rainfall patterns include not only torrential downpours during thunderstorms, but also the recently observed phenomenon of excessively heavy rainfall of short duration which nevertheless overwhelms the drainage systems e.g., waterways, drains, canals. In the Caribbean region flash flooding can occur within 1 to 2 hours of heavy rainfall in the small mountainous islands because of their size and the steepness of the gradients. Major contributing factors include the frequency and intensity of recent rainfall events and soil saturation levels.

2.1.2. Climate

Within the Caribbean region there are two seasons: the dry season which generally occurs from January to May and the wet season from June to December. As expected, climate influences the relationship between rainfall and runoff. During the dry season less than 1/3 of the annual rainfall occurs and during this period the soil becomes dry and increases its water holding potential (cracks and fissures may occur depending on the soil type). During the wet season the soil becomes saturated within weeks of the onset of the rains and much of the rainfall flows quickly over these saturated soils into water courses. These aspects of climate change and flooding are covered in Chapter 2.

2.1.3. Coastal Storms

Coastal storms often lead to flooding of low-lying coastlines. The combination of high tides and storms is likely to lead increasingly to flooding of coastal communities given the fact that sea-level s have risen by approximately 19centimetres (IPCC, 2013). The deposition of sea-water and action of powerful waves can potentially destroy coastal community infrastructure, contaminate coastal aquifers and result in saltwater intrusion of agricultural lands. On very rare occasions tsunamis occur in which large volumes of water are displaced in the ocean by the action of earthquakes or marine volcanic eruptions. The role of mangroves and wet lands in the prevention of floods and coastal erosion are outlined in Chapters 4 and 5.

2.1.4. Hurricanes

A tropical cyclone with winds of 64 knots or greater is called a hurricane in the Caribbean region. Hurricanes occur in various sizes and can range from several hundred metres to a thousand kilometres across. Hurricanes develop over warm tropical and subtropical waters and follow seasonal tracks. Early season hurricanes of June and July tend to begin in the Gulf of Mexico and western Caribbean while those of August, September and October tend to originate in a broad band stretching from Florida to the Cape Verde Islands. Most significant flooding occurs during hurricanes when considerable amounts of rainfall are added to already saturated soil. Impacts of flooding on the agricultural sector are described in

Chapter 3 and other related sectors including the Law relating to these man -made or natural disaster s are outlined in Chapters 4, 7 and 11.

2.1.5. Urban Storm Water Runoff

Urbanization has dramatically changed the drainage patterns of natural catchment areas by increasing the volume and rate of surface runoff. Very often the volume of water is within the carrying capacity of rivers prior to urbanisation but the runoff from intense rainfall cannot percolate through the soil due to built structures, and large surface area s covered with concrete and tarmac. In these circumstances the rainwater quickly exceeds the carrying capacity of drains, water courses and sewer systems creating torrents, which flood roads, basements and city centres (Figure 1). The impact of flooding on health is outlined in Chapter 6.

2.2. Artificial or Man-Made Causes of Flooding

2.2.1. Deforestation and Quarrying

Large areas of forest on mountain sides have been cleared in north Trinidad and in Jamaica largely to access areas for the quarrying of building materials, for urban housing and for road construction. The removal of the protective vegetation from the sides of mountains results in erosion of the top soil which finds its way into water courses and into the sea. The resultant mud flows move at a rapid rate and are usually more destructive than flood water alone. Mud slides destroy crops, flood streets, engulf homes and destroy infrastructure along their path due to the velocity of the moving wall of mud and water. In addition, numerous deaths have been reported due to flooding in the Caribbean region. Chapters 3, 4, 5 and 6 cover some of these issues in greater detail.

Figure 1.1. Urban flooding in Diego Martin, Trinidad during 2013 (Photo: Kristian De Silva, Republished courtesy the T &T Guardian).

2.2.2. Urbanization

The rapidly-growing global population and their need for housing have resulted in unprecedented land use changes, with the conversion of prime agricultural lands into housing areas.

This urban sprawl has forced farmers to advance into unsuitable lands, even very steep hillsides, destroying vegetation cover and drastically changing hydrological systems. In the Caribbean region such patterns are occurring at an alarming rate with the development of housing communities on the hills and mountain sides whether by legal means or by squatting, thereby increasing the risk of flooding. For example, a recent flooding episode which occurred in Diego Martin, Trinidad was blamed by disaster officials "on the denudation of hillsides by builders and 'slash and burn ' farming" (Figure 2).

As the demand for housing increases populations also move to geographic areas which are low lying and prone to flooding (see Chapter 7) but unfortunately many such householders report no insurance coverage and must rely on government hand-outs after flooding episodes.

2.2.3. Canals and Poor Maintenance

The construction of large canals, box drains and the desilting of major water courses is a major annual capital expenditure necessary to prevent flooding and for waste water management.

Unfortunately some of these structures are poorly constructed and maintained leading to their collapse during heavy rains – leading to flooding. Chapters 9 and 10 outline hard and soft engineering solutions to flood-prone areas with their wider application to the Caribbean region.

Figure 1.2. Flooding in Trinidad caused by the denudation of hillsides by builders and "slash and burn" farming. (Photo: Kristian De Silva, Republished courtesy the T &T Guardian).

2.2.4. Destruction of Wet Lands and Mangrove Habitats

An important ecosystem service provided by wet lands including mangrove ecosystems is in absorbing the water from rivers and slowing down the release of surface water into the surrounding oceans and seas.

These vegetation barriers including mangroves also protect the land from storm surges and from salt water intrusions into agricultural lands. Chapters 4, 5 and 6 outline the role of the coastal margins and wet lands in protecting small island states from e.g., sea-level rise, hurricanes and storm surges.

Figure 3 shows the destruction of the Godineau mangrove forest area in south Trinidad which is already causing flooding in areas which have had no history of flooding.

3. Flooding Physical Risk Factors in the Caribbean Region

Table 2 shows the physical risk factors associated with climate extremes which can affect Caribbean islands. Regional climate models raise significant questions regarding the amount of rainfall, reliability of rainfall and environmental responses to triggers of climate change. However, extreme events such as flooding will have a profound impact on sectors with close links to climate, such as water, agriculture and food security, forestry, health, and tourism. In fact there is *high confidence* that changes in the climate have the potential to severely affect water management systems. The challenge therefore is to find suitable adaptation strategies for both the short and long term - especially as small island developing states will be the first to be affected (Table 2).

Figure 1.3. Photograph of the destruction of the Godineau mangrove area in south Trinidad during 2013 (http://www.newsday.co.tt/day/1,61836.html).

Table 1.2. Caribbean island type and exposure to climate extremes

Island type	Exposure to climate risk
Plate-Boundary Islands large land areas high elevation high biodiversity well-developed soils river flood plains orographic rainfall	Jamaica and Trinidad are affected by flooding due to rivers and high elevations. Many housing areas are found on the steep slopes of the Northern Range (Trinidad) and Blue Mountains (Jamaica). Most major settlements are on the coast and exposed to storm surge and sea-level rise. In addition, many housing settlements are found within river flood plains (Caroni).
Inter–Plate (Oceanic) Islands-Volcanic islands steep slopes different stages of erosion barrier reefs relatively small land area less well developed river systems orographic rainfall	Most of the Windward and Leeward islands are small and exposed to hurricanes, which cause the most damage in coastal areas and housing areas. Barrier reefs often regulate storm surges and tsunamis.
Raised Limestone Islands steep outer slopes concave inner basins sharp karst topography narrow coastal plains no or little surface water no or minimum soil	Barbados, Carriacou and some Bahamas Islands experience storm surges and wave damage. These islands experience fresh water shortage and drought. Potable water problems can trigger health problems associated with water storage and poor soils may impede agricultural cultivation.

Adapted from Campbell 2006.

CONCLUSION

SIDs are vulnerable to the ravages of Climate Change and urgently require implementation of appropriate adaptation measures (Chapters 2 and 12). In addition, there is an urgent need for a 'common sense ' approach to water course management by clearing debris during the dry season (Chapter, 2), and using education programs to elicit community awareness and participation (Chapter 7). However, to ensure water courses remain clear, agencies should enforce appropriate legislation (Chapter 11).

REFERENCES

Allison EH, Perry AL, Badjeck MC, Adger NW, Brown K., Conway D, Halls AS, Pilling GM, Reynolds JD, Andrew NL, Dulvy NK. 2009. Vulnerability of national economies to the impacts of climate change on fisheries. *Fish and Fisheries*, 10:173-196.

Campbell, J. 2006. Traditional Disaster Reduction in Pacific Island Communities. *GNS Science Report 2006/38, Sciences*, Lower Hutt, New Zealand, 46 pp.

Church, J.A & White, N.J. 2006. A 20th century acceleration global sea-level rise. *Geophysical Review Letters* 33: L01602.

IPCC, 2007. Summary for Policymakers. In Climate Change 2007: The Physical Science Basis. *Contribution of Working Group I to the Fourth Assessment Report of the Intergovernmental Panel on Climate Chang*e (IPCC AR4) S. Solomon et al. eds (Cambridge University Press).

IPCC, 2012. Managing the Risks of Extreme Events and Disasters to Advance Climate Change Adaptation. *A Special Report of Working Groups I and II of the Intergovernmental Panel on Climate Change* [Field, C.B., V. Barros, T.F. Stocker, D. Qin, D.J. Dokken, K.L. Ebi, M.D. Mastrandrea, K.J. Mach, G.-K. Plattner, S.K. Allen, M. Tignor, and P.M. Midgley (eds.)]. Cambridge University Press, Cambridge, UK, and New York, NY, USA, 582 pp.

IPCC, 2013. Summary for Policymakers. In: Climate Change 2013: The Physical Science Basis. *Contribution of Working Group I to the Fifth Assessment Report of the Intergovernmental Panel on Climate Change* [Stocker,T.F., D. Qin, G.-K. Plattner, M. Tignor, S. K. Allen, J. Boschung, A. Nauels, Y. Xia, V. Bex and P.M. Midgley (eds.)]. Cambridge University Press, Cambridge, United Kingdom and New York, NY, USA.

Peterson, T. C., X. Zhang, M. Brunet-India, and J. L. Vázquez-Aguirre, 2008. Changes in North America n extremes derived from daily weather data. *J. Geophys. Res.,* 113, D07113, doi:10.1029/2007JD009453.

Polyak, L, Bischof, J, Ortiz, J.D, Darby, D.A., Channell, J.E.T., Xuand C., Kaufmane, D.S., Løvlie, R., Schneider, D.A., Eberl, D.D., Adler, R.E., Council, E.A. 2009. Late Quaternary stratigraphy and sedimentation patterns in the western Arctic Ocean. *Global and Planetary Change* 68: 5–17.

Sabine, C.L, Feely, R.A, Gruber,N, Key, R.M, Lee, K, Bullister, J.L, Wanninkhof, R, Wong, C.S, Wallace, D.W.R, Tilbrook, B, Millero, F.J, Peng, T.H, Kozyr, A, Ono, T, Rios, A.F. 2004. The oceanic sink for anthropogenic CO2. *Science,* 305 : 367-371.

In: Flooding and Climate Change
Editors: D. D. Chadee, J. M. Sutherland and J. B. Agard
ISBN: 978-1-62948-317-7

Chapter 2

FLOODING AND CLIMATE CHANGE IMPACTS IN THE CARIBBEAN REGION

Kishan Kumarsingh*
Ministry of the Environment and Water Resources,
Port of Spain, Trinidad, West Indies

ABSTRACT

Flooding issues in selected Small Island Developing States (SIDS) are examined; in particular those in the Caribbean, where data and information are available, and where the potential for flooding could result from projected climate change. Data show an increase in heavy rainfall events, including in places where total precipitation has decreased. In the Caribbean some areas of increased precipitation are noted for the months of December, January and February, while the region-wide decrease is enhanced in June, July and August, particularly around the Greater Antilles, where the model consensus is also strong. Future projections of precipitation changes for the Caribbean region shows a general annual mean decrease in precipitation across the entire region. Results from the HadCM3 model (Global Climate Models), downscaled for two scenarios, show a near-linear decrease in summer precipitation to the 2080s for a station in Jamaica. Downscaled results for Barbados and Trinidad, however, show increases rather than decreases.

While these results are indicative for a particular downscaled model and data set, it is important to note that there is no consensus across all models for changes in precipitation. Notwithstanding any uncertainty in changes in future precipitation, the unique and specific circumstances of small island states make them particularly vulnerable to these projected changes. In particular, the majority of socio-economic activity takes place on or near the coast in small island states. Management of risks, therefore, is of paramount importance in mitigating the impacts of future climate change in respect of increased flooding risks. The main management tools available for the future include: developing strategies for disaster risk management, sustainable land management, ecosystem management and ecological restoration and insurance to counter the expected increased risk of flooding impacts on SIDS may face in the future.

* E-mail: anamika_tt@yahoo.com.

1. INTRODUCTION

Flooding is usually caused by natural weather events, and may be exacerbated by anthropogenic influences. For example, heavy rainfall and thunderstorms, or intense precipitation over a short period; prolonged, extensive rainfall, all in combination with high tide conditions can be described as the natural causes of flooding in any given geographic area. However, other factors, in particular those that are man -made and generally within man's control, can also contribute to flooding severity e.g., poor maintenance of sewer and drainage networks and inadequate or poorly maintained watercourses. Land use and development issues are also factors that influence flooding, both in respect of severity and damage incurred as a result. For example, inappropriate development in flood plains; building on land in a way that prevents rainfall from draining away naturally, including paved areas such as highways, roads, car parks and recreational areas that are not only impermeable to water, but increase the risk of flooding from rainwater runoff ; and badly designed flood defence infrastructure. In this regard, land use change from natural vegetation such as deforestation in watershed catchments to built development causes greatest run-off of surface water and siltation of natural waterways. Natural geographic features such as elevation and distance from the coast (e.g., low lying coasts and small islands) may also determine ease and extent of flooding. Notwithstanding the synergistic effect s of natural and anthropogenic factors on flooding, in this chapter we focus on how climate variability and projected climate change can affect flooding and issues incidental thereto. Specifically, an examination of the following is undertaken:

- observed climate trends in respect of precipitation and sea level rise insofar as they affect inundation and flooding ;
- projected changes in precipitation with projected climate change under various scenarios with particular reference to the Caribbean region;
- potential impacts arising from these projections in the context of flooding ; and
- management of potential risks.

Flooding issues in selected Small Island Developing States (SIDS) are then examined; in particular those in the Caribbean, where data and information are available, and the potential flooding that may result from projected climate change.

2. OBSERVED CLIMATE TRENDS AND FUTURE PROJECTIONS - PRECIPITATION AND SEA LEVEL RISE

2.1. The Intergovernmental Panel on Climate Change Fourth Assessment Report

The Intergovernmental Panel on Climate Change (IPCC) (2007) in their Fourth Assessment Report stated that observations since 1970 indicate that precipitation has generally increased over land north of 30°N over the period 1900 to 2005 but downward trends tend to be dominant over this 105 years. From 1900 to the 1950s, precipitation increased significantly in the area 10°N to 30°N, but decreased after circa 1970. In the deep tropics from 10°N to 10°S, there has been an observed decrease in precipitation especially after 1976/1977. However, it is noteworthy that observations of significant precipitation changes, where they do occur, are consistent with measured changes in streamflow.

The IPCC (2007) report further stated that

> "... changes are occurring in the amount, intensity, frequency and type of precipitation. These aspects of precipitation generally exhibit large natural variability, and El Niño and changes in atmospheric circulation patterns such as the North Atlantic Oscillation have a substantial influence. Pronounced long-term trends from 1900 to 2005 have been observed in precipitation amount in some places: significantly wetter in eastern North and South America, northern Europe and northern and central Asia, but drier in the Sahel, southern Africa, the Mediterranean and southern Asia. In northern regions more precipitation now falls as rain rather than snow. Widespread increases in heavy precipitation events have been observed, even in places where total amounts have decreased. These changes are associated with increased water vapour in the atmosphere arising from the warming of the world's oceans, especially at lower latitudes. There are also increases in some regions in the occurrences of both droughts and floods ", and that.... "substantial increases are found in heavy precipitation events. It is likely that there have been increases in the number of heavy precipitation events (e.g., 95th percentile) within many land regions, even in those where there has been a reduction in total precipitation amount, consistent with a warming climate and observed significant increasing amounts of water vapour in the atmosphere. Increases have also been reported for rarer precipitation events (1 in 50 year return period), but only a few regions have sufficient data to assess such trends reliably".

Observed trends in sea level rise as reported by Bindoff et al. (2007) are) summarized in Figure 1. They have assessed the rate of sea level rise for 1961 to 2003 as 1.8 ± 0.5 mm yr^{-1} and for the 20th century as 1.7 ± 0.5 mm yr^{-1}. Sea level is determined by the thermal expansion due to sea surface warming as well as from contributions due to the melting of ice sheets.

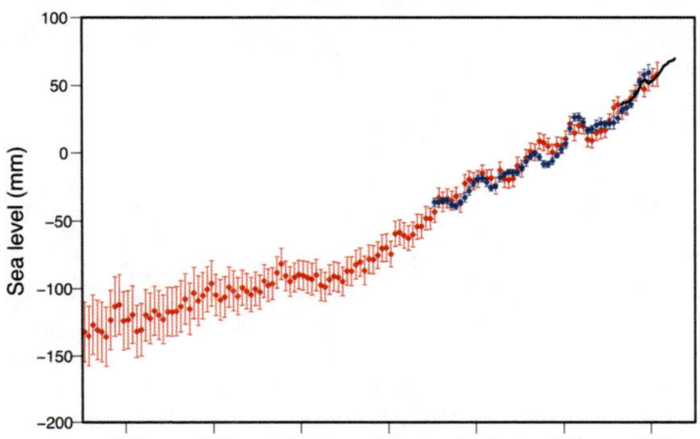

Figure 2.1. Annual averages of the global mean sea level (mm). The red curve shows reconstructed sea level fields since 1870; the blue curve shows coastal tide gauge measurements since 1950, and the black curve is based on satellite altimetry as presented in the IPCC Fourth Assessment Report (2007).

Projected changes in precipitation and sea level rise are based on various future scenarios. Without going into too much detail on the development and meaning of the scenarios, it is sufficient to say that the scenarios are based on possible future biophysical and socio-economic conditions, (including concentrations of greenhouse gas es in the atmosphere, technology development, and changes in socio-economic behavior) on which models of changes in the future climate are based and for different future time periods. These models are generally referred to as General Circulation Models or Global Climate Models (GCMs), which are sometimes statistically downscaled to bring it to a finer resolution, particularly for smaller land masses such as SIDS.

Globally, the IPCC (2007 see Parry et al., 2007) has estimated that based on models ; mean global temperature is expected to increase, as are mean water vapour, evaporation and precipitation. There is also a projected increase in precipitation in those regions that have tropical precipitation maxima (e.g., monsoons) and over the tropical Pacific in particular. However, it is projected that there would be general decreases in the subtropics, and increases at high latitudes as a consequence of a general intensification of the global hydrological cycle. Future projections of precipitation changes for the Caribbean region are summarised in Figure 2, which shows a general annual mean decrease in precipitation across the entire region.

Some areas of increased precipitation are noted for the months of December, January and February, while the region-wide decrease is enhanced in June, July and August, particularly around the Greater Antilles, where the model consensus is also strong. Results from the HadCM3 model (a GCM), downscaled for two scenarios, show a near-linear decrease in summer precipitation to the 2080s for a station in Jamaica. Downscaled results for Barbados and Trinidad, however, show increases rather than decreases. While these results are indicative for a particular downscaled model and data set, it is important to note that there is no consensus across all models for changes in precipitation.

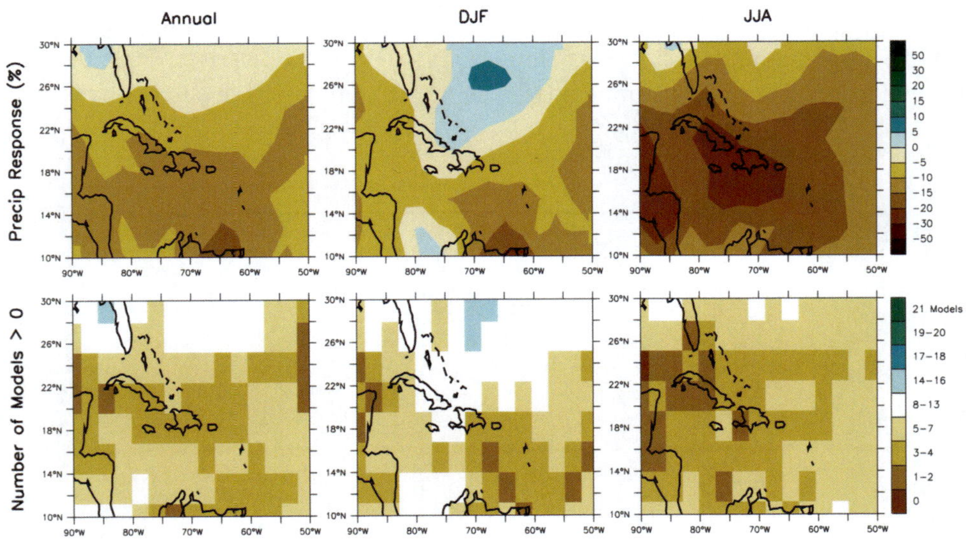

Figure 2.2. Precipitation changes over the Caribbean Top row: Annual mean, December, January, February and June, July, August fractional precipitation change between 1980 to 1999 and 2080 to 2099, averaged over 21 models. Bottom row: number of models out of 21 that project increases in precipitation (From IPCC 2012).

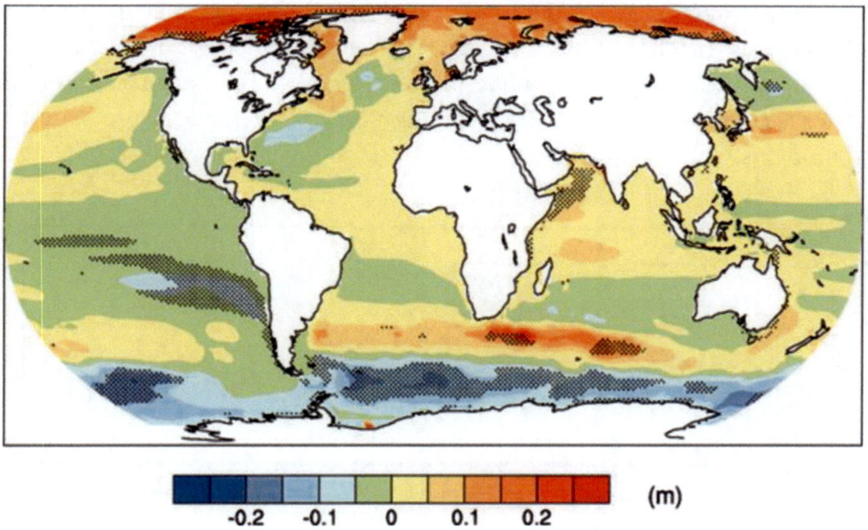

Figure 2.3. Local sea level change (m) due to ocean density and circulation change relative to the global average (i.e., positive values indicate greater local sea level change than global) during the 21st century, calculated as the difference between averages for 2080 to 2099 and 1980 to 1999, as an ensemble mean over 16 GCMs forced with one future scenario. Stippling denotes regions where the magnitude of the multi-model ensemble mean divided by the multi-model standard deviation exceeds 1.0. (From IPCC 2012).

In respect of sea level rise, the IPCC estimates an increase in sea level within the range 0.18 to 0.59 m to the year 2100. However, these changes will not be uniform across the globe as a result of changes in ocean circulation and density as shown in Figure 3. It is to be noted here as well that this Figure is a synoptic indication based on a single future scenario and therefore does not provide absolute values.

2.2. Recent Scientific Reports

Since the publication of the IPCC Fourth Assessment report in 2007, there has been a number of scientific publications providing up-to-date data on precipitation and sea level rise. For example Levitus et al., (2012) have shown a strong positive linear trend in world ocean heat content since 1955, with one third of the observed warming occurring in the 700-2000 m layer of the ocean (Figure 4), thereby indicating higher than previously estimated sea level rise potential.

Observed sea level s from satellite altimetry (Meyssignac and Cazenave 2012) indicate that the global mean sea level rose at rates of 1.77 mm/yr and 3.2 mm/yr during the 20th century and the last two decades respectively, as a result of both an increase of ocean thermal expansion and land ice loss (Figure 5). Additionally, there is evidence that islands have suffered significant sea level rise since 1950 (Becker et al., 2012). Projections of sea level rise have also been estimated to be higher than those estimated in the IPCC Fourth Assessment Report (2007). Other studies (Verimeer and Rahmstorf 2009; Nicholls and Cazenave 2010) have estimated a global sea level rise ranging from 75 to 190 cm for the period 1990–2100.

In a new study (Durack et al., 2012) analyzing ocean salinity found consistency with earlier model-based suggestions that dry regions will become drier and wet regions will become wetter in response to warming (Giorgi et al., 2011) and that the water cycle has intensified twice as much as that simulated by climate models over the same period. Furthermore, it has been suggested that the more extreme rainfall amounts closely follow the rainfall intensity (Benestad et al., 2012), implying that we can expect more extremes in both floods and droughts.

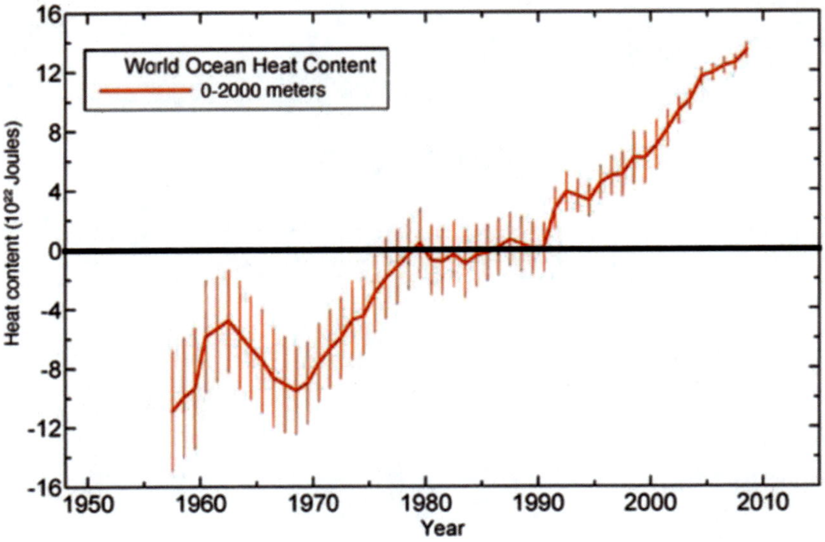

Figure 2.4. World ocean heat content and thermosteric sea level change (0-2000m) from 1955 to 2010. Adapted from Levitus et al. (2012), Geophysical Research Letters at the time of writing. (*Courtesy Bill Hare and Climate Analytics - Survive Project*).

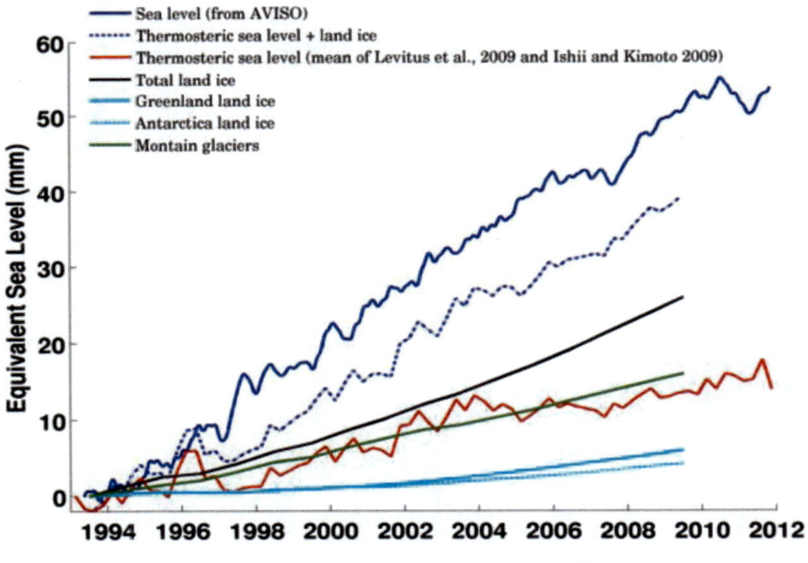

Figure 2.5. Observed sea level from satellite altimetry over 1993-2010 (blue solid curve); thermal expansion (red curve); contribution from Greenland and Antarctica (cyan curves) and glaciers (green curve). The black curve represents the total land ice contribution while the blue dotted curve represents the total climatic contribution (sum of thermal expansion and land ice). Adopted from Meyssignac and Cazenave (2012), Levitus et al. (2009), Ishii and Kimoto (2009) (*Courtesy Bill Hare and Climate Analytics - Survive Project*).

The IPCC (2012) Special Report on managing the risks of extreme events and disasters to advance climate change adaptation (SREX) has indicated that the frequency of heavy precipitation or the proportion of total rainfall from heavy falls will likely increase, and future flood losses in many locations will increase in the absence of additional protection measures. In respect of tropical cyclones, the IPCC SREX has projected that the global frequency of tropical cyclones will likely either decrease or remain essentially unchanged. However, a recent study (Bender et al., 2010) using an operational hurricane-prediction model that produces realistic distribution of intense hurricane activity for present-day conditions, predicted a nearly doubling of the frequency of category 4 and 5 Atlantic storms by the end of the 21st century.

The IPCC (2012) SREX report has also indicated that it is very likely that mean sea level rise will contribute to upward trends in extreme coastal high-water levels in the future.

3. Potential Impacts from Projected Changes and Management Approaches

The IPCC (2007) has estimated that *"coasts are projected to be exposed to increasing risks, including coastal erosion, due to climate change and sea-level rise. The effect will be exacerbated by increasing human -induced pressures on coastal areas"*. They further assert that "*many millions more people are projected to be flooded every year due to sea-level rise by the 2080s. Those densely-populated and low-lying areas where adaptive capacity is relatively low, and which already face other challenges such as tropical storms or local coastal subsidence, are especially at risk. The numbers affected will be largest in the mega-deltas of Asia and Africa while small islands are especially vulnerable..." and...."adaptation for coasts will be more challenging in developing countries than in developed countries, due to constraints on adaptive capacity"*.

The IPCC (2012) SREX indicates that it is likely that the frequency of heavy precipitation or the proportion of total rainfall from heavy rainfalls will increase in the 21st century over many areas of the globe. The very likely contribution of mean sea level rise to increased extreme coastal high water levels, coupled with the likely increase in tropical cyclone maximum wind speed, is a specific issue for tropical small island states. Notwithstanding any uncertainty in changes in future precipitation, the unique and specific circumstances of small island states make them particularly vulnerable to these projected changes.

In particular, the majority of socio-economic activity takes place on or near the coast in small island states. Coastal flooding due to increases in storm surge as a result of increased intensities of tropical cyclones provides a significant risk to the economic and social well-being of island states.

Management of risks therefore is of paramount importance in mitigating the impacts of future climate change in respect of increased flooding risks. It is not the intention in this chapter to analyse the merits or demerits of management approaches to mitigating future climate change impacts but merely to highlight the various approaches that can be taken given the peculiar nature of the circumstances to which they can be applied. In this regard, several approaches are identified. Those identified in the IPCC SREX (Cutter et al., 2012) in the context of flooding include:

- *developing strategies for disaster risk management* in the context of climate change requires a range of approaches, informed by and customized to specific local circumstances given that disasters, including flooding occurs largely at a localised level;
- *sustainable land management* is an effective disaster risk reduction tool, and includes land use, land use planning, zoning for specific uses, conservation and preservation zones, buffer zones, as well as land acquisition for such purposes. Small island states are faced with this specific challenge given their limited land space as well as the need for development, often resulting from political and economic pressures. However, such measures can prove to be more sustainable at the local level than structural measures.
- *ecosystem management and ecological restoration* activities that focus on addressing deteriorating environmental conditions are critical approaches given the invaluable ecosystem services that they can provide. Such activities include, among others, watershed rehabilitation, agro-ecology, and forest landscape restoration, for example, sustainable agricultural practices that exclude slash-and-burn on hillsides, a practice that is prevalent in some Caribbean islands, can go a long way in mitigating floods.
- *insurance* is a risk transfer mechanism that can be used at the local level to minimize costs and encourage recovery

4. Selected Case Studies in Small Island States (Caribbean and, Where Available, Pacific and IMA Region)

The following case studies illustrate the impacts of climate -related events on selected small island developing states in the context of flooding risks. Future climate change and climate variability are projected to increase the intensity of these extreme events and so increase the potential risk of flooding impacts that island states may face in the future.

4.1. Caribbean

According to the Caribbean Disaster Emergency Management Agency (http://weready.org/flood/index.php), flooding in the Caribbean occurs most frequently during the hurricane season, which lasts from June 1st to November 30th, and is attributed to the heavy rains that accompany these extreme events. Not all Caribbean countries are affected equally as the extent of flooding is largely dependent on topography, level of development as it relates to infrastructure, and whether they have been affected by an intense hurricane or weather system.

Historically, countries such as Haiti, Dominican Republic, Barbados, Antigua and Barbuda, St. Vincent and the Grenadines, the British Virgin Islands, Bermuda and Belize, Jamaica and St. Lucia have been most affected by flooding.

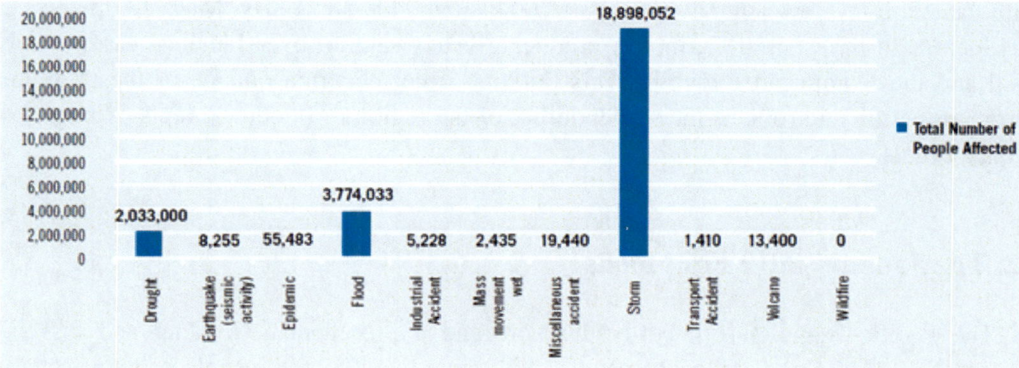

Figure 2.6. Total number of people affected by different types of disasters in the Caribbean (1980-2009) (From UNDP 2011).

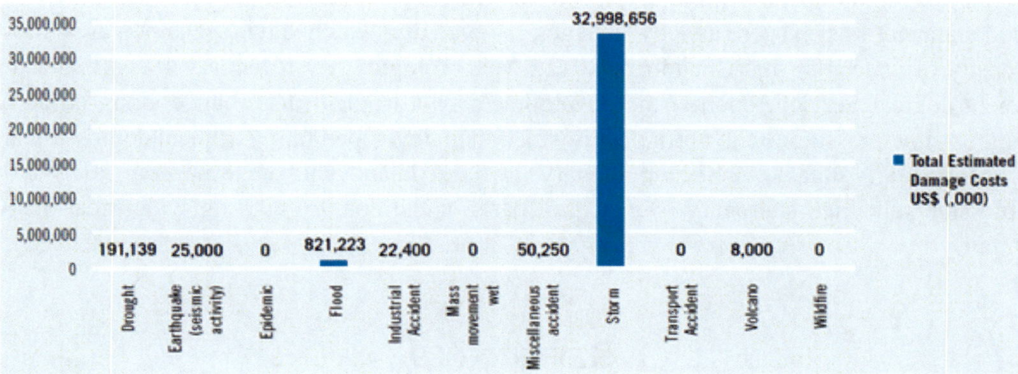

Figure 2.7. Total estimated damage costs in US dollars for different types of disasters in the Caribbean (1980-2009) (From UNDP 2011).

Figure 6 illustrates that climate -related events affected the majority of persons in the Caribbean during the period 1980 -2009 (UNDP 2011). This is also reflected in Figure 7 which shows the associated damage costs (UNDP 2011).

5. Case Studies

5.1. The May 2004 Caribbean Floods

The May 2004 flooding occurred in the Caribbean from May 18 to May 25 2004 and affected the Caribbean islands, mainly Haiti, the Dominican Republic, and some parts of northern Puerto Rico (Flood Archive 2004), and caused severe damage in the Dominican Republic and Haiti, with over 1,300 homes being destroyed and about 2,000 people being killed, caused mainly by landslides and drowning (NOAA 2004). Significant rainfall accompanied the storm, with over 9.7 inches of rain falling in Haiti, and 10 inches falling in the Dominican Republic (NOAA 2004). Meteorologically, the heavy rainfall started with the development of a broad low pressure system over Central America that was accompanied by heavy rainfall on May 19 (NHC 2012), which drifted into the Caribbean Sea (http://australiasevereweather.com/cyclone/2004/summ0405.htm.). By May 23 the system was located in the Central Caribbean producing heavy rainfall over Jamaica, eastern Cuba, Haiti and the Dominican Republic. This flooding event occurred outside of the traditional hurricane season and reflects the susceptibility of the region to flooding events as a result of climate events.

5.2. The January 2012 Fiji Floods

The Pacific island of Fiji suffered severe and unprecedented flooding and losses in January 2012. The floods killed 11 people, with more than 319,000 people being affected in low-lying areas. Fiji officials estimated that 55 per cent of export crops, mostly sugar cane, were destroyed, with the estimated damage costs at around $F30.6 million ($US17 million) (UNDP 2011, http://www.wsws.org/articles/2012). Significantly, it is also suggested that flood damage was exacerbated by changes in land use which have seen increases in the intensity of water run-off (Chaudhary 2012). Meteorologically, on January 19, 2012 the Fiji Meteorological Service reported the development of a tropical disturbance along an active trough of low pressure to the north of Fiji which then developed into a tropical depression and moved across Vanua Levu during January 21. (Fiji Island Climate Summary 2012). The depression remained stationary over central parts of the country causing widespread heavy rainfall.

REFERENCES

Becker, M., B. Meyssignac, C. Letetrel, W. Llovel, A. Cazenave and T. Delcropix. 2012. Sea level variations at tropical Pacific islands since 1950. *Global and Planetary Change*. V. 80-81. pp85-98.

Bender, M.A., T.R. Knutson, R.E. Tuleya, J.J. Sirutis, G.A. Vecchi, S.T. Garner and I.M. Held. 2010. "Modeled Impact of Anthropogenic Warming on the Frequency of Intense Atlantic Hurricanes." *Science;* 454-458.

Benestad, R.E., D. Nychka, and L.O. Mearns 2012. "Spatially and temporally consistent prediction of heavy precipitation from mean values", *Nature Climate Change* 2: 544-547.

Bindoff, N.L., J. Willebrand, V. Artale, A, Cazenave, J. Gregory, S. Gulev, K. Hanawa, C. Le Quéré, S. Levitus, Y. Nojiri, C.K. Shum, L.D. Talley and A. Unnikrishnan, 2007: Observations: Oceanic Climate Change and Sea Level. In: Climate Change 2007. The Physical Science Basis. *Contribution of Working Group I to the Fourth Assessment Report of the Intergovernmental Panel on Climate Change* [Solomon, S., D. Qin, M. Manning, Z. Chen, M. Marquis, K.B. Averyt, M. Tignor and H.L. Miller (eds.)]. Cambridge University Press, Cambridge, United Kingdom and New York, NY, USA.

Chaudhary, F. 2012. *"Fiji flood damage exacerbated by land use changes."* http://pidp. eastwestcenter.org/pireport/2012/February/02-03-13.htm (Accessed June 13 2012).

Cutter, S., B. Osman-Elasha, J. Campbell, S.-M. Cheong, S. McCormick, R. Pulwarty, S. Supratid, and G. Ziervogel, 2012. Managing the risks from climate extremes at the local level. In: Managing the Risks of Extreme Events and Disasters to Advance Climate

Change Adaptation [Field, C.B., V. Barros, T.F. Stocker, D. Qin, D.J. Dokken, K.L. Ebi, M.D. Mastrandrea, K.J. Mach, G.-K. Plattner, S.K. Allen, M. Tignor, and P.M. Midgley (eds.)]. *A Special Report of Working Groups I and II of the Intergovernmental Panel on Climate Change* (IPCC). Cambridge University Press, Cambridge, UK, and New York, NY, USA, pp. 291-338.

Durack, P.J., S.E. Wijffels, and R.J. Matear, 2012. "Ocean Salinities Reveal Strong Global Water Cycle Intensification During 1950 to 2000", *Science* 336: 455-458.

FICS 2012. *Fiji Islands Climate Summary January* 2012 Volume 33 Issue 1.

Flood Archive 2004.. http://www.dartmouth.edu/~floods /Archives/2004sum.htm (Accessed June 13 2012).

Giorgi, F., E. Im, E. Coppola, N.S. Diffenbaugh, X.J. Gao, L. Mariotti, and Y. Shi, 2011. "Higher Hydroclimatic Intensity with Global Warming", *Journal of Climate*, 24: 5309-5324.

http://australiasevereweather.com/cyclones /2004/summ0405.htm. Monthly Global Tropical Cyclone Summary May 2004.

http://weready.org/flood/index.php?option=com_content&view=article&id=16&Itemid=17 (accessed June 13 2012).

http://www.wsws.org/articles/2012/apr2012/fiji-a11.shtml (Accessed June 13 2012).

IPCC 2007. *Contribution of Working Group II to the Fourth Assessment Report of the Intergovernmental Panel on Climate Change,* 2007. M.L. Parry, O.F. Canziani, J.P. Palutikof, P.J. van der Linden and C.E. Hanson (eds) Cambridge University Press, Cambridge, United Kingdom and New York, NY, USA.

IPCC, 2007. Summary for Policymakers. In: Climate Change 2007: Impacts, Adaptation and Vulnerability. *Contribution of Working Group II to the Fourth Assessment Report of the Intergovernmental Panel on Climate Change*, M.L. Parry, O.F. Canziani, J.P. Palutikof, P.J. van der Linden and C.E. Hanson, Eds., Cambridge University Press, Cambridge, UK, 7-22.

IPCC, 2012. Managing the Risks of Extreme Events and Disasters to Advance Climate Change Adaptation. *Special Report of Working Groups I and II of the Intergovernmental Panel on Climate Change* [Field, C.B., V. Barros, T.F. Stocker, D. Qin, D.J. Dokken, K.L. Ebi, M.D. Mastrandrea, K.J. Mach, G.-K. Plattner, S.K. Allen, M. Tignor, and P.M. Midgley (eds.)]. Cambridge University Press, Cambridge, UK, and New York, NY, USA, 582 pp.

Ishii M, Kimoto M. 2009. Reevaluation of historical ocean heat content variations with varying XBT and MBT depth bias corrections. *J. Oceanogr.* 65:287–99.

Levitus, S., J. I.Antonov, T. P. Boyer, R. A. Locarnini, H. E. Garcia, and A. V. Mishonov 2009. Global ocean heat content 1955–2008 in light of recently revealed instrumentation problems, *Geophys. Res. Lett.,* 36.

Levitus, S., J.I Antonov, T.P Boyer, O.K Baranova, H.E. Garcia, R.A. Locarnini, A.V. Misonov, J. Reagan, D. Seidov, E.S. Yarosh, and M.M Zweng. 2012. World ocean heat content and thermosteric sea level change (0-2000), 1955-2010, *Geophys. Res. Lett.*, 39: 10 doi:10.1029/2012GL051106.

Meyssig nac, B. and A. Cazenave 2012. Sea level: A review of present-day and recent-past changes and variability. *Journal of Geodynamics* 58: 96-109.

Nicholls, R.J. and A. Cazenave (2010). Sea level rise and its impact on coastal zone. *Science.* 328: 1517-1520.

NOAA 2004. *Special Tropical Disturbance Statement (Report).* National Hurricane Center. http://www.nhc.noaa.gov/archive/text/DSAAT/2004/DSAAT.200405191247 (Accessed June 13 2012).

NOAA 2012. *Global Hazards and Significant Events: May 2004*". National Oceanic and Atmospheric Administration. http://www.ncdc.noaa.gov/sotc/hazards /2004 (Accessed June 13 2012).

UNDP 2011. Caribbean Implementation of the Hyogo Framework for Action. *Mid-Term Review.* 2011. UNDP Caribbean Risk Management Initiative.

Vermeer M. and S. Rahmstorf, 2009, "*Global sea level linked to global temperature*", doi10.1073pnas.0907765106.

In: Flooding and Climate Change
Editors: D. D. Chadee, J. M. Sutherland and J. B. Agard
ISBN: 978-1-62948-317-7

Chapter 3

IMPACT OF FLOODING ON THE AGRICULTURAL SECTOR

Adrian Cashman*
Centre for Resource Management and Environmental Studies (CERMES),
The University of the West Indies, Cave Hill Campus, Barbados, West Indies

ABSTRACT

This chapter outlines the problem of how the world will feed 9 billion people by 2050. The effect flooding may have on this problem is first addressed with an exploration of the relationship between agriculture and water via the early development of agriculture in river flood plains, the development of irrigation, and the sophisticated early water management schemes such as were operated by the Chinese and by the Maya in Central America. Examples are given of recent crop losses due to flooding in the Caribbean. How agricultural land is managed can have a role in flood mitigation and adaptation. The present land-use patterns in the Caribbean developed primarily as a result of historical demand for tropical export crops rather than reflecting the local soil, topography, natural ecosystems and climate. It is suggested that the impact of flooding on agriculture can be viewed using a Source-Pathway-Receptor approach. Appropriate land management and agricultural practices that retain water in the landscape (source control measures) contribute to flood-risk mitigation and could provide longer warning times. There has been greater interest in food security in the Caribbean since the spike in food prices in 2008. Greater production of local foodstuffs would require better management of land, water and nutrients and higher use-efficiency of these resources. Suggested measures include improved rainwater harvesting through the use of pools and small storage dams. Payment for Ecosystem services (PES) schemes are of growing interest for potential rehabilitation, conservation and maintenance of watersheds, the best-known example being that of FONAG in Ecuador. It is also suggested (FAO, 2010) that increased use of Agroforestry can help to meet the triple challenge of improving food security, mitigation, and reducing the vulnerability and increasing the adaptability of agricultural systems to climate change by diminishing the effects of extreme weather events.

---

* Corresponding author: Phone: 1246-417-4316; Fax: 1246 424 4204; Email: Adrian.cashman@cavehill.uwi.edu.

1. INTRODUCTION

It has been estimated that by 2050 the global population will have risen to 9 billion people. Of the current 7 billion people one billion are at risk of starvation everyday so if we as a planet are already struggling to meet the current basic food requirements then how is agricultural production going to be able to meet the future challenge of an extra two billion mouths to feed every day. As societies become more affluent and incomes rise, as more developing countries transition to being developed, diets and preferences change. The most commonly cited example is that of China where there has been an increase in demand for fruit, milk, poultry products and meat products in tandem with investment in the expansion and modernisation of China's food processing, distribution and retail networks (Dong and Fuller, 2007). In the past, increases in population and the concomitant demand for agricultural products would have been responded to by, among other strategies, increasing the area of land under cultivation. Conversely, declines in population, such as occurred during the 14th century in Europe led to the abandonment of land and a waning of agricultural activity. In the Caribbean region the issue of how to feed an additional two billion people is overshadowed by a more immediate concern. Since the 1960s many countries have moved from a situation of having a substantial agricultural trade surplus to a deficit with food imports competing for importance with those of oil -based products (Deep Ford et al., 2007). The Region now imports more agricultural products than it exports, and the area or land under cultivation has decreased in most of the insular Caribbean (UNEP/CARICOM, 2004). How to address food security and the role of agriculture has been a growing concern at the national and regional level, which has led to proposals such as the Jagdeo Initiative by the former President of Guyana to achieve a resurgence of regional agricultural potential.

Agriculture relies on water to be able to grow the variety of produce that we consume as food ; use for fibre; produce fuel, pharmaceuticals and materials. Rainfed agriculture and natural ecosystems, as a generalisation, use most of the available renewable water resources, through the uptake of soil moisture and evapotranspiration by plants, so called green water (Falkenmark and Rockström, 2004, see Figure 3.1). On top of this irrigation uses a significant amount of blue water (from rivers, lakes and groundwater). Green water flows take about 65% of global precipitation with the rest being blue water. Of the total blue water flows about 9% is used for agriculture whilst 11% of green water flows support crop and livestock production (Falkenmark and Rockstrom, 2006). This use by agriculture is said to be around 70-75% of the world's water resources, in comparison domestic and industrial use are much less significant though no less important. About 20% of what is used by agriculture comes from groundwater, though this depends on the part of the world being considered. Given the significant usage of water by irrigated agriculture, and the location of many irrigated areas in flood plains it is clear that there is a close relationship between, agriculture, water management and by extension, flooding. Flooding's relationship to agriculture may be likened to a Goldilocks effect; we like it when it is just right but too little or too much can prove to be disastrous.

Globally, there are more people living in urban than in rural areas, it is in these rural areas that the majority of agricultural activities take place. In the Caribbean the rural population accounts for 33% of the overall population, though of course it is higher in some countries than in others. Furthermore, rural areas experience much higher rates of poverty than urban areas, globally the figure is given as 70% and over 80% of rural households rely on farming as their source of income to some extent, and typically it is the poorest households that rely on farming and farm labour the most (UNDP, 2006). In many Caribbean countries at least 20% of the working population depend on agriculture for their livelihood and, importantly, have limited opportunities to shift to other income-earning sectors. Limited opportunities together with other conditions of rural poverty often result in rural households being more vulnerable and less resilient to the effects of shock events as they have fewer assets to fall back on. Shock events can include disasters associated with extreme weather events; events that threaten household livelihoods and income generating activities. The risk environment in which rural agricultural households operate has in recent times become riskier, new factors such as natural resource degradation and climate change have to be factored in (IFAD, 2011). Being able to manage or avoid risks through mitigation and adaptation measures is a prerequisite for moving out of poverty.

Agriculture, through its reliance on the services nature provides and especially the inter-relationship with climatic conditions is particularly vulnerable in the tropics to natural environmental hazards such as tropical cyclones, droughts, high temperatures and flooding. Climatic variability and the seasonality of water availability exercise a limiting influence on agricultural productivity and exacerbate the problems associated with achieving food security. Globally the numbers of natural disaster s being reported has increased dramatically since the 1950s (Figure 3. 2) with much of is the apparent increase due to better communication and improved reporting. The number of people killed through disasters, as a percentage of global population the numbers of people affected, has increased by a factor of two. Climate-related disasters have shown a much greater increase than geophysical disasters with the most common climate -related disaster being flooding. Furthermore, climate-related disasters are thought to account for some 90% of disaster- related displacements of people. The direct and indirect economic costs associated with disaster events are huge, and the burden of these costs falls most heavily on developing countries. With respect to flooding, most of the damage is associated with urban areas so that agriculture bears only an estimated 1-5% of the total damage cost of flood-related events – in developed countries. In countries in which agriculture plays a significant role in the economy and where the numbers of people dependent on the agricultural sector for their livelihoods is high the impact of climate-related disasters, and in particular flooding, may be much greater.

Agriculture plays an extremely important role in the Caribbean both in terms of contribution to local economies and its potential role in poverty alleviation. The challenges that Caribbean States are facing and how they are responded to should be seen as contributing to the evaluation of options that the global community has in responding to feeding a growing population in the face of increasingly challenging environmental conditions. In the following sections the physical and historical bases of agriculture will be reviewed, together with an overview of the agricultural characteristics of agriculture within the Caribbean. This will be followed by looking at the causes of flooding and the impacts that flooding can have on agriculture, with particular reference to recent Caribbean experience. Finally, the question is addressed of how the impacts of flooding on agricultural activities might be alleviated through mitigation and adaptation measures.

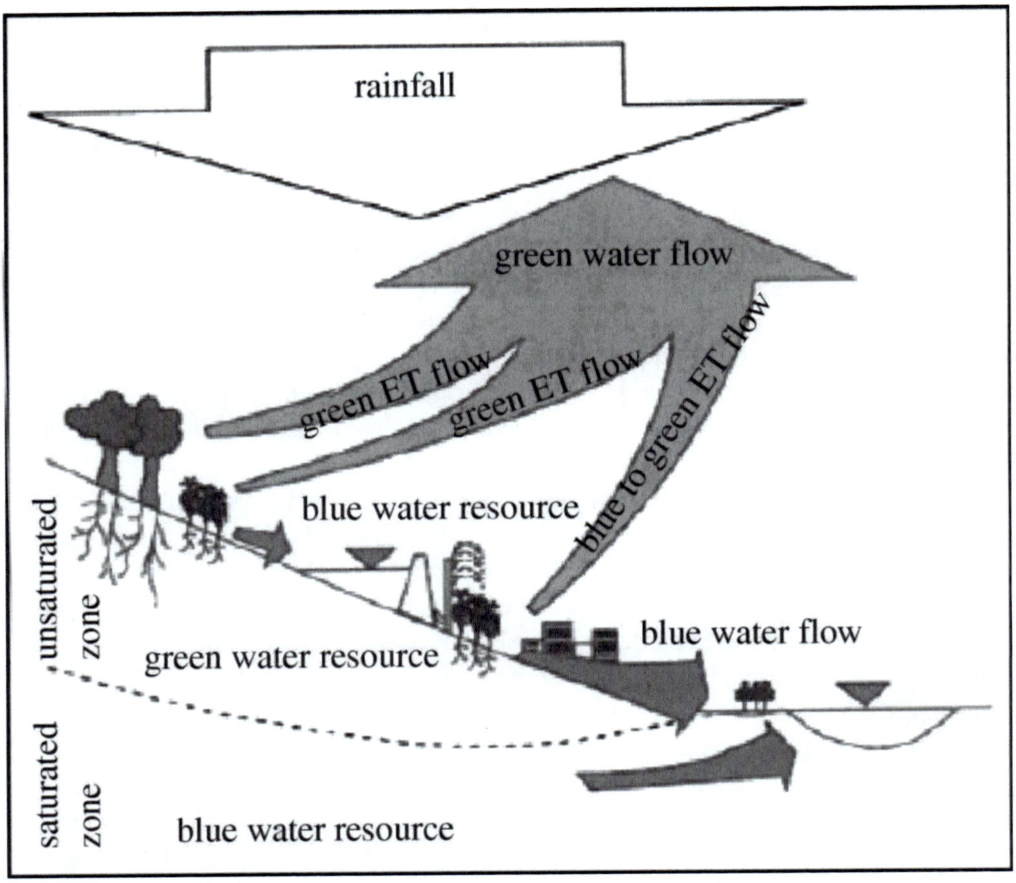

Source: Falkenmark and Rockstrom, 2006.

Figure 3.1. Blue and Green Water Flows.

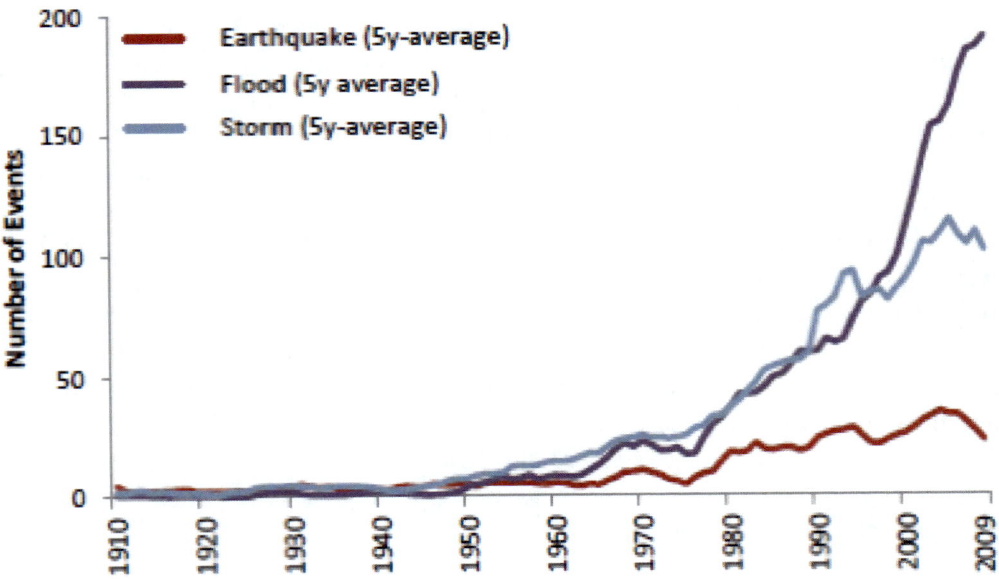

Figure 3.2. Number of Disaster Events by year, 1910-2009 (WEF, 2011).

2. DIRT

In a BBC radio parody of gardeners, one of the characters replied to any question with "I think the answer lies in the soil", as good a place as any to start. After all soil and land are an essential part of agriculture. At its simplest, soil formation takes place as a result of interactions between air, water, rocks, chemicals and biotic activity. Water percolating through soil provides the driving force for soil formation, through either physical and/or chemical weathering. With physical weathering rock may be broken down but the chemical composition remains essentially the same. In the case of chemical weathering, water's mildly acidic composition as it enters the ground enables chemical reactions to take place resulting in the decomposition of rocks, changes in chemical composition and physical properties, and the making available of minerals. The availability of minerals and moisture enable plant life to establish itself which in turn can accelerate the rate of soil formation. It does so through respiration by roots, which raises levels of carbon dioxide in the soil, and through the formation of humic acid from decaying organic matter (Figure 3.3).

Soil helps to complete life cycles by providing a medium through which decomposition and recycling of organic material takes place. This helps to filter and convert dead material into nutrients in the soil and living organisms in soil are responsible for the processes of nutrient cycling and mixing of materials, all of which improve soil texture. Soils make nutrients available for further use by other plants and organisms, which allow oxygen to be supplied as well as retained. It therefore follows that the availability of nutrients can constrain the productivity of terrestrial ecosystems that rely on the soil as a medium for their processes.

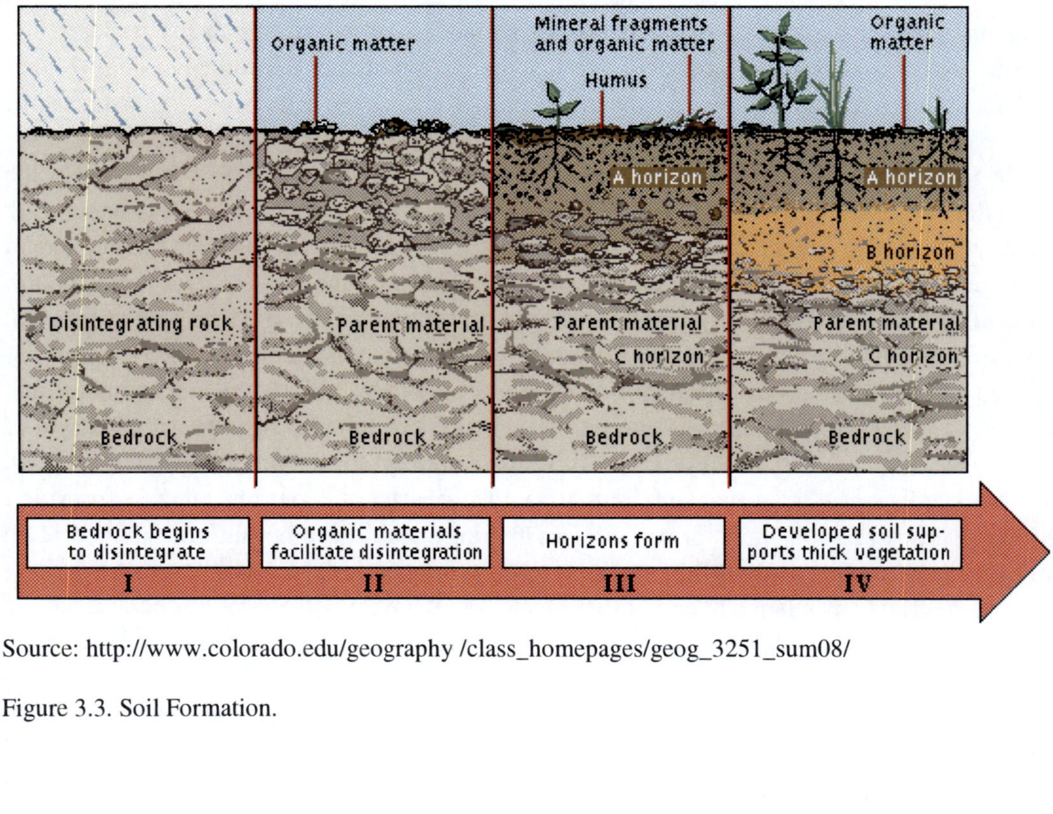

Source: http://www.colorado.edu/geography /class_homepages/geog_3251_sum08/

Figure 3.3. Soil Formation.

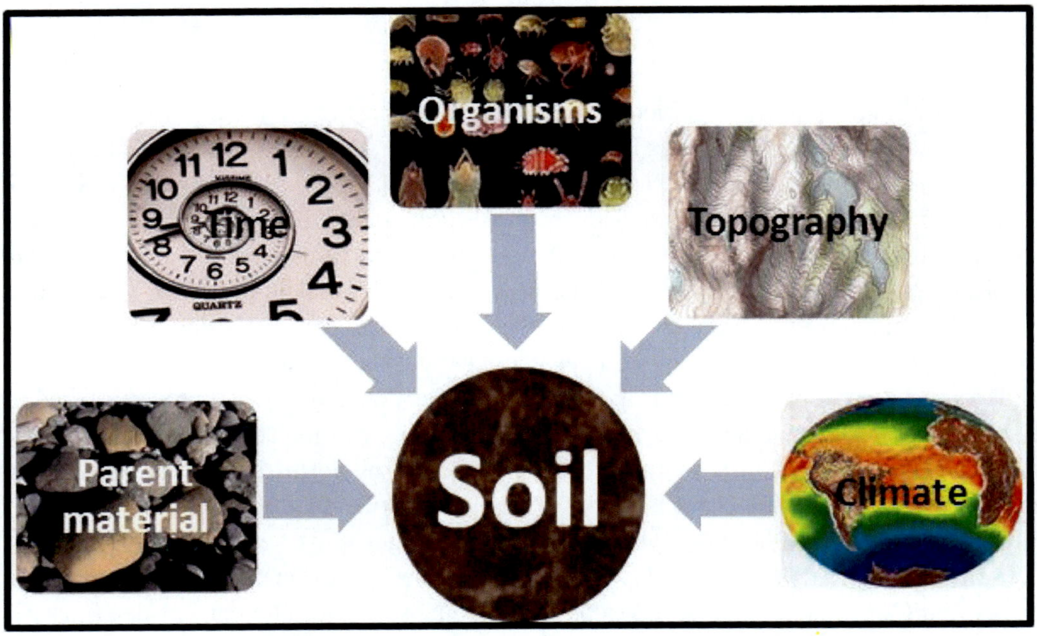

Figure 3.4. Factors affecting soil formation.

All this can take a very long time, a fully developed soil can take up to 1000 years to form. Other things being equal, at some point in time the rate of soil formation will come into equilibrium with other environmental processes such as erosion. This suggests that soil has a characteristic thickness that is determined by environmental circumstances operating within a particular landscape ; it is governed by the interaction of biotic organisms, topography, climate, parent material and time, evolving together as a mutually dependent system (Figure 3.4). Landscape geomorphology has the effect of modulating the effects of organisms and of climate on soil formation, so steep slopes will lead to thinner soils especially in humid areas due to increased levels of run-off, whilst low-lying areas in the same area can lead to the accumulation of organic matter. The geomorphology of an area often reflects the underlying geology which also has a major influence on the type of soil material, as does the geological processes of weathering in situ and erosional processes (alluvial, colluvial or aeolian). Climate through the effects of precipitation and temperature affects the rates of the physical processes of soil formation as well as the rates of biological and chemical activity in soils. It also has an indirect effect through its influence on biotic activity; plants, animals and organisms. In addition to processes present in the soil there can also be external processes of disturbance that can help to physically recycle material; tree falls, fire, erosion. Topography, climate, parent material, organisms and time together act as controls on soil formation through processes of transformation (weathering and decomposition), translocation (movements within the soil profile), addition (organic matter or deposition) and loss (leaching or erosion). Thus any change in one of these factors will have an effect on the character of the soil, which is exactly what human activity does when it seeks to harness soils for its own productive purposes.

Erosion processes mirror those of soil formation and are governed by the same parameters, indeed erosion is sometimes part of soil formation processes. Rates of soil erosion are strongly influenced and controlled by topography (principally slopes) and by the presence or absence of agricultural activities and their influence on vegetative cover. The major erosional process of interest here is that involving the flow of water. Where sufficient precipitation occurs on a land surface, water in the form of run-off can accumulate and the ability of water to transport material is determined by the rate of run-off. The higher the speed of run-off, the greater the erosional capacity of the water and its ability to transport material away. Flowing water carves out channels and widens existing drainage paths which can themselves increase the erosion caused by flowing water by exposing more soils surfaces. On steep slopes intense rainfall can saturate soils, which not only means that more water cannot enter the soil but has to flow over it, but also can cause landslides or mudflows. A lack of vegetation cover increases the erosion capability of flowing water by sometimes as much as by 100 times, which is well above the natural rates of soil formation. Thus agricultural practices which leave soil exposed, often for parts of the year, contribute to the increased erosion of soils, affecting the very resource that allows farmers to engage in productive agricultural activities. Moreover cultivation of soils often leads to a decline in soil organic matter so that the soils themselves become more erodible the longer they are cultivated, if remedial measures are not taken.

The Caribbean is located in the tropics, an area of generally high though seasonal rainfall and warm temperatures. As a result, in its undisturbed state, much of the region was covered by tropical forests, with the steady, moist trade winds interacting with soils and the topography of the islands to determine of the various ecosystems present (Lugo et al.,1981). In their natural condition, only a small proportion of rainfall in these forests reaches the forest floor as direct throughfall or flow down tree trunks and stems (Figure 3.5). Most precipitation reaches the ground through drips from the canopy with the reset evaporating off from the canopy leaves and branches. Of the precipitation that reaches the ground, the majority infiltrates into the soil, through leaf litter Critchley and Bruijnzeel, 1997). Leaf litter protects the ground against the erosive effects of the impact of raindrop splash on the soil surface as well as of surface run-off. Furthermore, the undisturbed soil usually has better infiltration properties and water -holding capacity, which helps limit surface run-off thus minimising erosion. Tropical forests support large amounts of plant biomass even though soil fertility is low; part of the reason for this is the relatively closed nutrient cycle. Plant nutrients derived from leaf litter, aerosols, and dust are cycled between the canopy and the soil, with only small amounts leaking away. This is aided by forests having extensive surface root system s that are capable of intercepting and trapping most of the nutrients entering the system, thus offsetting the leaching of minerals and soluble constituents out of the soil that is typical of high rainfall and temperature conditions. This helps to explain why soils that develop under tropical, forested conditions rapidly lose their fertility and erosion rates can be very high.

As has been made clear, soil formation depends on the interplay of five key variables which over time tend towards a meta-stable state in which soil formation and soil erosion achieve a degree of balance. Any significant change in conditions will not only disturb the balance but also over a period of time change the characteristics and properties of the soil. Change also brings other unintended consequences, with soils being less resilient and more susceptible to the adverse effects of natural processes such as the effects of heavy and prolonged rain. These tend to act as positive feedback mechanisms with consequences including increased rates of erosion, loss of fertility and impoverishment of soils leading to reduced fertility and yields. Indeed it has been speculated that the economic collapse that characterised the period between the 1880's – 1940's in the Caribbean was "caused in part by ecological constraints resulting from short-sighted exploitation of forest lands" (Lugo et al., 1981).

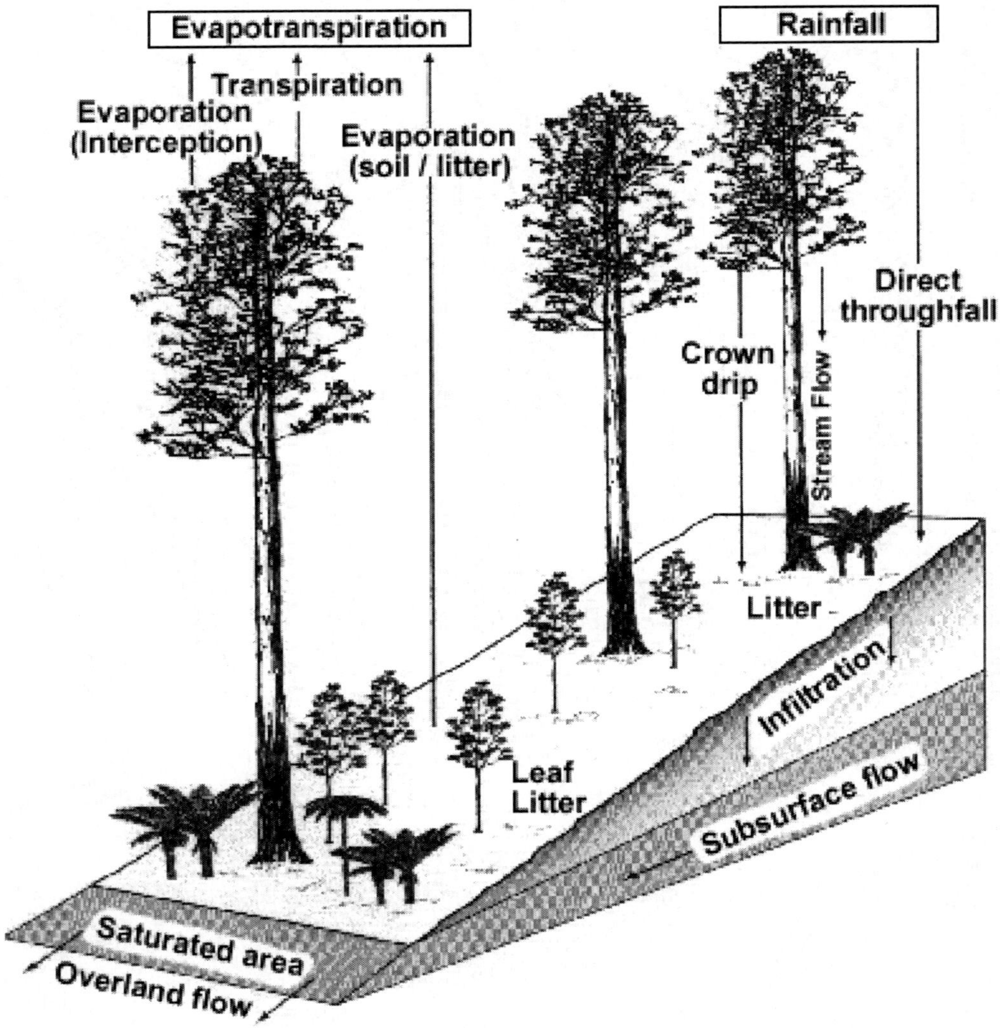

Source: Predicting Water Yields From Mountain Ash Forest Catchments' Industry Report of the Co-operative Research Centre for Catchment Hydrology.

Figure 3.5. Forest hydrological cycle.

3. Hydraulic Civilisations and the Rise of Agriculture

Whilst flooding is usually taken as being synonymous with negative impacts it should not be forgotten that it was what enabled the expansion of agriculture that led to the production of food surpluses and provided the basis for the rise of urban civilisation. The imitation of flood events gave rise to the development of irrigation and the controlled application of water for productive purposes. This to a large extent freed agriculture from its reliance on weather, bringing areas under cultivation that would otherwise be unsuitable, increasing yields and extending growing seasons. Without flooding and the harnessing of seasonal flood waters none of this would have been possible. Flood agriculture arose at different places and at different times around the globe, the staple crops grown varied, but where it did take place it provided the foundation for what have been called 'hydraulic civilisations'. These are so-called because their existence was dependent on a mastery of hydraulics and the ability to harness and deliver floodwaters through networks of structures such as canals, sluices, reservoirs, and dykes to control flooding and drainage to prevent water-logging in flat, poorly drained soils.

Some of the earliest examples of irrigated agriculture developed in semi-arid areas alongside large rivers that flooded seasonally bringing on the floodwaters quantities of silt that covered the land as the flood waters receded. These were areas where rainfall was insufficient to support rain-fed agriculture. The benefits of having access to a reliable, year-round water supply, and fertile silt more than offset the disadvantages of such areas and the considerable efforts required to harness and control the water resources to make them available for agricultural production. The challenge of how to exploit seasonal floods and their attendant benefits was met through the growth of centralised planning and authoritarian organisation. Maximisation of agricultural potential required the development of a specialised body of knowledge, and the ability to mobilise and direct mass labour to construct and maintain the hydraulic infrastructure. In such societies power and hierarchical social organisation complement one another, and in turn, depend on regimented and comprehensive control of water supply. In China, so closely entwined were river management and the organisation and exercise of power that the Chinese character for politics, three drops of water next to a dyke, is derived from the root word meaning flood control. The building and maintenance of irrigation enabling water works, wherever it occurred, led to an increase in agricultural productivity, the creation of food surpluses and the ability to support higher population numbers.

Whilst flood irrigation brings with it the benefits of a more reliable food supply, there are drawbacks, which, by a twist of fate, are associated with the long-term effects of irrigation. Irrigating field crops in areas of high groundwater and high rates of evaporation, brings dissolved salts to the surface through capillary action. Irrigation, especially flood irrigation, keeps the groundwater level high and sustained irrigation adds salts dissolved in the irrigation water to the soil in a mutually reinforcing cycle. While natural flushing from periodic flood events can act as a control, in poorly drained soils this is not enough to control the salt build up. Overtime the once fertile soils become salinised, lose their fertility and agricultural productivity declines. To compensate more and more marginal land is brought under cultivation but often to no avail. If the irrigation infrastructure is not well maintained, leading to silting up of drainage ditches then the system declines even faster. The history of irrigation is full of examples of the adverse effects of prolonged flood irrigation on soils that were not managed with a view to minimizing the build up of salts: the Harappan civilisation on the Indus River; and the Sumerian civilisation on the lower reaches of the Rivers Tigris and Euphrates. Indeed, in Mesopotamia the trend over time was for irrigated agriculture to move up-river in response to problems experienced with managing the soils, and the gradual loss of agricultural productivity. Given natural climatic variability it is inevitable that there would be very large flood events, which would overwhelm the irrigation infrastructure. The memory of such a catastrophic flood probably gave rise to the flood story known as the Epic of Gilgamesh from Sumeria and the later Biblical Story of Noah and the Great Flood.

In China, extensive cultivation and the development of irrigation took place far inland, more than 1000 km from the sea, along the Yellow River from around the third millennium BC. The silt-laden floodwaters deposited their loads providing the raw material for agriculture. But the Yellow River is prone to exceptional floods which burst river banks, wash away infrastructure and farmlands, with disastrous consequences and causing millions of deaths. Measures to protect land against flooding by the Yellow River has been a constant concern, and for more than 4000 years there has been a major official charged with flood control. The ploughing of increasing areas of easily erodible soils, over time, increased the silt loads carried downstream by the river. Such was the effect on the river that its name changed from the Great River to the Yellow River in the fourth century BC.

Source www.latinamericanstudies.org/aztec-society.htm.

Figure 3.6. Aztec chinapas.

Some of the earliest evidence in the Americas for the harnessing of floodwater comes from the middle Zaña Valley in Peru where canals, garden plots and domestic sites have been found dating to around 3400 BC (Dillehay et al., 2005). The canals follow the contours of the valley with intakes that intercept seasonal flood waters as well as a perennial stream allowing a lower terrace bench to be farmed, allowing food production to support increases in population. In Central America the large urban civilisations such as the Olmecs, Aztecs and Maya would not have been able to support the large population numbers without the ability to produce major food crops. In the highlands of Mexico irrigation would have been necessary for agriculture of any extent, whilst the challenging environment of the Maya heartlands would also have required a flexible response. Archaeological and ethnographic work over the last few decades has pointed to the importance of the utilisation of wetlands and their modification to support the production of food surpluses (Baker, 2007). These areas show evidence of canals built between wetlands to divert water to create new farmland as well as evidence of raised fields (chinampas) separated by ditches which provided both water and drainage (Figure 3.6). In addition, there is evidence of extensive terracing on slopes with check dams and other features to control run-off whilst maintaining soil moisture.

It is clear from this brief overview that the advent of irrigated agriculture and its supporting infrastructure had far reaching effects. It created the ability to support large urban and rural population numbers, and at the same time required the stratification of societies and the differentiation of work. It required the development of tools of control and social mobilisation for the continued provision of surpluses brought about by the provision of water on a regular and continuous basis, by means of flooding of fields, to crops that would otherwise not flourish. Equally successful irrigation required the capture and utilisation of flood waters whilst at the same time creating the need for protection against floods that would otherwise damage or destroy it. A delicate balancing act between too much and too little, the consequences of either could be devastating (Figure 3.7).

from Gómez-Pampa et al., 1982, "Experiences in traditional hydraulic agriculture " in Maya subsistence, ed: Kent V. Flannery, Academic Press. p. 327-342.

Figure 3.7. Raised garden beds.

4. Flooding

Rainfall will, given the right combination of circumstances, give rise to flooding. In the Caribbean the type of rainfall events that cause this are often associated with tropical storms or hurricanes. In other words the flooding that occurs is in many instances accompanied by high, gale-force winds so that the damage that arises is caused by a combination of wind and flooding and it can be difficult, when assessing the impact, to distinguish between the two. ECLAC (2010a) reported that between 1990 and 2008, there were 165 natural disaster s, of which 61% were windstorm-related events – hurricanes, floods and tropical storms, with flooding related to the after effects of hurricanes estimated to have caused 29% of the natural disasters. Of the 14 countries studied in the ECLAC report (2010a) Haiti, Dominican Republic and Jamaica experienced the greatest number of storms and flood related events, and for Belize, Dominica, the Dominican Republic, Guyana, Haiti, Jamaica, Saint Lucia and Suriname, the agricultural sector was most affected by windstorm-related natural disasters (Table 3.1). In these countries the direct damage and losses accounted for between 80% and 85% of the total national impact of natural disasters. Country-level assessments of the losses and damage caused by natural disasters are undertaken after extreme events and in the Caribbean mostly use the methodology developed by ECLAC (ECLAC, 2003). However, such assessments are undertaken only after major events, with the result that the number and impact of smaller events are often missed.

Table 3.1. Natural disasters by type and country: 1990 – 2008

Country	Drought	Earthquake	Epidemic	Extreme Temperature	Flood	Mass movement (wet)	Storm	Total
Belize				1	3		8	12
Guyana	1				4	1		6
Suriname					2			2
Anguilla							1	1
Bahamas							12	12
Cayman Islands							7	7
Dominica		1					5	6
Dominican Republic		1	3		13		16	34
Grenada							4	4
Haiti	3		1		22		23	49
Jamaica	1		2		4		14	21
Netherlands Antilles							1	1
St Lucia		1				1	4	6
Turks and Caicos							5	5
Total	5	3	6	1	48	2	100	165

Source: EM-DAT: The OFDA/CRED International Disasters Database www.emdat.be

When talking of flooding it is useful to distinguish between run-off and inundation as they are often conflated. Run-off, and particularly stormwater run-off, occurs when water falling on the ground cannot infiltrate into the soil, if there is a slope then this water makes its way to the lowest point. Stormwater run-off will flow either overland as a sheet flow or collect in channels, streams and rivers. Aspects such as the duration and intensity of the rainfall event, the infiltration capacity of the soil and its degree of saturation, the nature of the topography, as well as other soil factors already mentioned all have an influence in determining the severity of the stormwater run-off. Flow velocity is proportional to its carrying capacity - as velocity increases so the erosional capacity of the flowing water increases, as does the volume of eroded material it is able to convey. Thus the ability to control stormwater run-off is critical to managing its erosional capacity. At the same time stormwater run-off also represents a valuable water resource which can be harnessed for beneficial use. Although structures, whether they are natural ones such as stream channels or man -made, only have a limited capacity to handle floodwaters, and it is a fact that, at certain times, this capacity will be exceeded, which often exacerbates flooding problems. In the case of rivers and streams, they will overflow their banks and spread into the surrounding areas, this often gives rise to the formation of flood plains, especially in the lower reaches of rivers.

Inundation means being covered with water, especially floodwaters but it has the additional suggestion that the water has collected in a particular place. It will remain there until it is either able to infiltrate into the soil and underlying formations, evaporate or otherwise be drained away. Depending on timing, inundation and flood recession can result in ideal conditions for the planting and growing of crops but the wrong timing or prolonged inundation can lead to crop losses. Thus, as with stormwater run-off and flows, it is when there is 'too much' that flooding changes from being an enabling factor to being a problem. What constitutes too much depends on timing during the growing cycle, the frequency, depth and duration of the event.

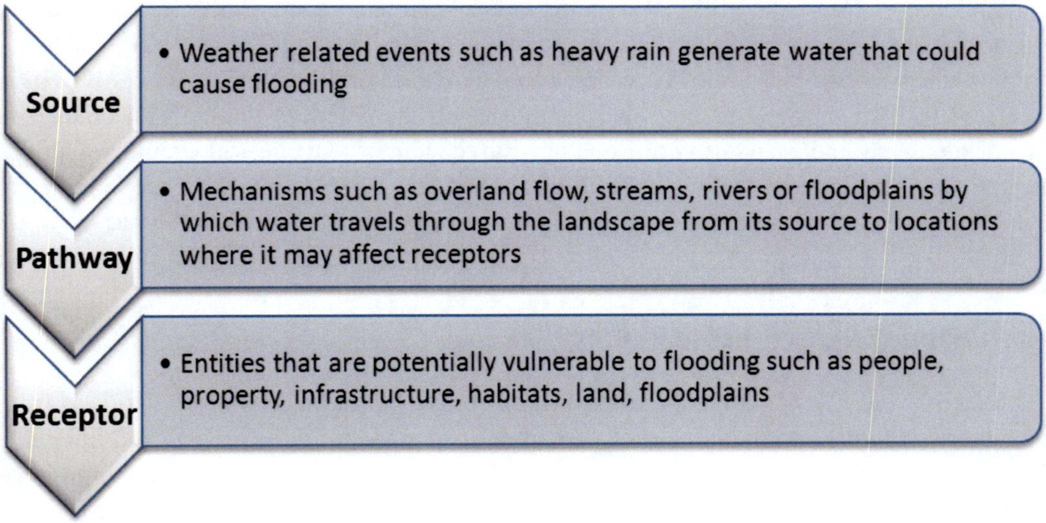

Figure 3.8. Source-Pathway-Receptor Model of flood risk (Morris et al., 2010).

Flooding is a natural phenomenon on which ecosystems have evolved to make use of. Flooding only becomes a 'risk' when people and property are co-located where flooding takes place. The consequences of flooding therefore depend on the vulnerability of the people and assets that are exposed, and the degree of exposure, or probability of experiencing a flood. One way of thinking about the impact of flooding on agriculture is to use a *Source-Pathway-Receptor* approach, see Figure 3.8 (Morris et al., 2010). A flood event has a cause or source, for example an extreme event such as a tropical storm or hurricane which is accompanied by very heavy rainfall : waters which have the potential to give rise to flooding are conveyed through pathways such as channels, streams and rivers and they at some stage reach a Receptor, where flooding occurs. If there is a disconnect at any point then no flooding occurs. The risks associated with the flooding are dependent on the consequences associated with the flooding and the probability of a flood (of a particular magnitude) occurring. Risks may be managed either through mitigation measures such as modifications to the source and/or pathway that act to reduce the probability of a particular event occurring. Alternatively, adaptation measures can be implemented which try to reduce the impact of flooding in the receptor areas.

Given that agriculture not only occupies significant areas of land, but also that it often occupy areas in the upper catchments as well as middle and lower reaches, it can have a significant role in flood mitigation and adaptation. Thus the way land is managed and the type of agricultural practices can have a very important impact on the severity of flooding arising from all but the most severe or extreme weather-related events. Poor soil structure and the degree to which surfaces become compacted due to repeated cultivation can lead to excess run-off from areas utilised for agriculture, whether for crops or livestock. The way land is managed can also influence the nature of the flow pathways, their connectivity, and hence the speed with which water travels across or through the landscape. Whilst the presence of features such as farm tracks, plough lines and the like, that can channel and concentrate water flows, are important, there is little evidence that shows that, under extreme events, such features actually exacerbate flooding at the catchment scale (O'Connell et al., 2007). The evidence seems to suggest that land management practices have the most effect on small to medium events. The widely held view is that appropriate land management and agricultural practices that retain water in the landscape, sometimes referred to as 'source control measures ', contribute to flood risk mitigation and could provide longer warning times. Adaptation measures involve changes in land use and management practices that can accommodate greater frequency and/or intensity of flooding with the aim of reducing vulnerability. Options include switching to crops that can tolerate longer periods of inundation, switching away from arable crops or modifying the type of agriculture by switching to low- or no-till practices.

4.1. Examples of the Impact of Flooding

Exactly what sort of mitigation or adaptation measures can be considered will vary from one place to another and on the type of agriculture being practiced. In Guyana and Suriname large scale agriculture is practiced on relatively flat, low-lying coastal areas adjacent to large rivers. Agriculture has developed based on the control of water through the landscape by means of extensive systems of dykes, channels and sluices. In these countries agricultural areas are very much receptors and rely on the management of flow to irrigate fields. Here flooding is an integral part of agriculture. The flooding is more associated with slow onset events, rainfall in the interior which gradually makes its way to the coastal areas through river flows. Measures such as proper maintenance of the flood control infrastructure, adequate flood warning, the provision of flood resilient infrastructure and the adoption of flood-tolerant crops, all provide ways to minimise the risks associated with flooding. It is very seldom that flooding is associated with extreme rainfall events in the coastal areas, though this does happened, the 2005 floods in Guyana being an example. Guyana and Suriname differ from the rest of the Caribbean in that they are not as affected by tropical storm- or hurricane-related natural disaster s.

During the 2005 floods in Guyana some 290,000 people were affected by the flooding and, of these, 92,000 had their homes flooded. The estimated damage to the agricultural sector arising from the severe floods that occurred in 2005 in Guyana amounted to the equivalent of 8% of the country's GDP, and the total loss associated with agriculture was 10% of the overall total. By contrast with the 2005 floods where the effects of prolonged, persistent localised rainfall led to flooding that affected wide sections of society, the 2006 and 2008 floods affected the agricultural sector. These floods were triggered by similar weather patterns, heavy rainfall in the interior which commenced several weeks in advance of the actual flood events. The prolonged rainfall before the floods allowed an increased soil moisture content and a reduced infiltration capacity thus contributing to increased overland flow and flooding. The total loss from the 2006 floods was approximated at USD30.1 million which was 4.8% of the GDP (ECLAC, 2006) of which the agriculture sector bore 74% of the impact. In 2008, over 45000 acres of rice was affected and the total losses in the agriculture sector were approximately USD6 million. Although there were many causes for the floods such as unusually heavy and prolonged rainfall and high antecedent soil moisture conditions in the case of the 2006 and 2008 floods, reports indicate that human agency was also a contributory factor, certainly in the case of the 2005 floods, with the failure to properly maintain the flood protection and management infrastructure being part of the problem.

It must be remembered that the coastal agricultural areas of Guyana and Suriname are located in floodplains which are of high agricultural value and it is therefore no wonder that they are prone to flooding. In such areas the impact of flood (stormwater) run-off is not so severe, but the water logging of soils as well as soil erosion will be significant for most of these farmlands and farmers. Of particular concern in these areas is the potential combination of sea-level rise and on-going land subsidence due to drainage controls. These areas are already below mean sea level and thus vulnerable to storm surges, riverine floods or a combination of the two. As receptor areas, the management and maintenance of flood protection infrastructure together with adaptation measures to reduce vulnerability are important.

In other parts of the Caribbean, the impact on agriculture from floods has been just as serious. Just a few examples will illustrate this point.

In Trinidad major flooding is associated with the lower lying, agricultural areas and it has been estimated that over the past 20 years it has caused damage to agricultural produce worth over US$20 million and caused damage to over 1,000 homes and businesses (Ramlal and Baban, 2008). Most of the flooding occurs during the rainy season and is associated with high intensity rainfall events that lead to flash floods, rather than the longer term inundation experienced in Guyana. The increasing occurrence of flooding in Trinidad has been ascribed, by some, to the clearing of previously forested areas for housing development as well as forest fire s and shifting hillside cultivation, especially in the Northern Range hills (Ramlal and Baban, 2008). Whilst these developments have increased agricultural flooding they have also increased urban flooding as well.

Jamaica is prone to flooding associated with tropical storms which regularly impact the island. In 2010 Tropical Storm Nicole caused damage to the road network, drainage systems, river protection and associated infrastructure valued at US$123 million, and the Ministry of Agriculture and Fisheries estimated damage to the agricultural sector of US$11.6 million. In addition damage to the water infrastructure was estimated at US$3.1 million.

Figure 3.9. The Mahaut market under water, Dominica.

Figure 3.10. Flooding on the Layou Park Road, Dominica.

In the same year Tropical Storm Tomas had a severe effect on St Lucia as it followed a major drought, the calculated return period for the rainfall was 1:180 years, whilst the combined probability of a drought followed by the rainfall event was estimated to be in excess of 1:1,000 years (ECLAC, 2011). The preceding drought conditions set the stage for extremely high potential for surface erosion and mass movements on slopes in the event of just a normal rainy period. The estimated damage and losses to the agricultural sector amounted to some 16% of the total losses, with the banana subsector being the most severely affected with whole fields being destroyed by the flooding. The plants that remained produced significantly reduced yields and in addition, the fields required major investments in clearing, silt removal, fertilization and drainage to restore harvests levels. Some 2000 farmers were estimated to have been affected through the loss of their crops and land used for cultivation (ECLAC, 2011).

Dominica is an interesting case; it is a mountainous island and has over 60% of forest cover, mostly concentrated in the higher elevations in the interior, though there has been some forest conversion mostly in the lower elevations where lands have been given over to agriculture. There has also been a gradual transition from large scale agriculture to small farms, which has had implications for land conservation measures. Small farmers have tended to cultivate their full acreage with short-term crops in order to maximise their financial returns. In 2011 the country was hit by a number of natural disaster s between July and September. In July heavy rains resulted in major flooding damaging infrastructure in the main agricultural area in the interior. In September more heavy rains resulted in landslides and then later in the month Tropical Storm Ophelia caused flooding (see Figures 3.9 and 3.10). Much of the brunt of the damage was borne by the agricultural sector of the country, which employs 25% of the labour force and generates 15% of GDP. The estimated cost of the damage and losses was USD30 million which amounts to 6.5% of GDP (IMF, 2012). The breach of the Matthieu Dam that took place as a result of the July rains is expected to have a long-term adverse impact on agriculture in the area.

5. Food Security

Although generalisations are dangerous, we might say that there are two types of farm holding that dominate in the Caribbean. On the one hand there are small family farms most of which have a limited area of land, often of under 2 hectares which, on the smaller islands can be found occupying mountainous areas. Though exact figures are scarce and often out of date small family farms account for the vast majority of farms through the region. By contrast the other type of farm holding is what might be referred to as plantations, which are significant in terms of land area and production (Roseboom et al., 2001). The development of large-scale agribusinesses is, in most cases, the exception rather than the rule, though countries such as Belize, Guyana and Suriname which have relatively large areas of land plus good water resources, are moving to develop their agriculture. Another feature is the relative specialisation in one or two major crops, usually as major export crops. The present land use patterns developed primarily as a result of historical demand for tropical export crops rather than reflecting the characteristics of soil, topography, natural ecosystems and climate. Both large and small farm holdings are susceptible to land degradation : large-scale agriculture because of the predominance of intensive monoculture; and small holdings because they can often be located on marginal lands, susceptible to erosion. In the greater part of the English speaking Caribbean, the amount of irrigated land is between 5 – 25% of the land available for cropping and again in the overwhelming majority of cases, cropland accounts for less than 10% of the land area. This can be ascribed to the mountainous terrain of many of the countries, which makes much of the land unsuitable for cultivation; in the wider Caribbean only Suriname, Haiti, Guyana, Dominican Republic and Cuba have significant areas of irrigated land in excess of 100,000 hectares.

Since the spike in world food prices in 2008 greater interest has been shown towards the issue of food security in the Caribbean. With some notable exceptions, the Caribbean is a net food importer and the trade gap between food imports and exports grows wider each year. There is also a dependence on a narrow range of food commodities, making the region highly vulnerable to food insecurity (Deep Ford, et al., 2007). There is a recognition that trade liberalisation and open economies will not result in increased food security for the Caribbean, unless there is much greater participation and expansion of agricultural production by local producers. Which, given the limitations faced by Caribbean agriculture, is unlikely except in particular cases. Thus a food self reliance strategy that depends on world markets is unlikely to result in food security for the region. The alternative is greater food self-sufficiency through greater production of local foodstuffs.

However, the agricultural sector is not well-placed to respond to this challenge except in a few cases. Some commentators have said that more productive and resilient agriculture requires the better management of land, water and nutrients, and a much higher use efficiency of these resources and inputs to agricultural production (FAO, 2010). Many of the measures proposed to achieve this would also have concomitant benefits in the field of flood water management (UNEP/CARICOM, 2004). Suggested measures include improved rainwater harvesting and retention of water in the landscape through the use of pools, small storage dams, and the management of soil fertility and organic matter which not only improves yields but also soil structure. In the longer term alternative farming strategies such as minimising tillage or the adoption of agroforestry are needed. According to the FAO (2010) agroforestry can help to meet the triple challenge of securing food security, mitigation and reducing vulnerability and increasing the adaptability of agricultural systems to climate change by diminishing the effects of extreme weather events.

6. Flood Risk Management Measures

The last few examples of flooding illustrate the importance of flood pathways (how water flows through the landscape) and their impact on agriculture. It is clear from these examples that overland- as well as flow-through drainage networks (streams and rivers) play a major role in flooding. This is especially so in the insular Caribbean where a combination of mountainous topography, land conversion and a significant amount of agricultural activity on slopes compound the flood risk . The examples also illustrate the point made earlier that although, in the case of small to medium climatic events, mitigation and adaptation measures can be adopted that will assist in the reduction of flood risk and damage, when it comes to major events, such as tropical storms attention should focus on post-event support measures which will aid recovery and lessen social and financial impacts. In this respect, timely flood warnings are one measure that can be adopted. Examples of this exist for example in Jamaica, which has developed a community monitored flood warning system (see Chapter 10), the Caribbean Institute for Meteorology and Hydrology (CIMH) developed a flood forecasting tool that gives a 48 hour weather prediction, and has applied this to Haiti. Other measures have included the development of flood risk maps though because of the cost and effort involved these have only been done for a very small number of places through the Caribbean. In addition to these measures the role of insurance as a post-flooding recovery vehicle has been discussed but due to the costs involved, for individual farmers, this is not thought to be a viable option. However, the Caribbean Catastrophic Risk Insurance Facility (CCRIF) is a regional fund available to governments to enable them to purchase coverage for earthquakes, hurricanes and excess rainfall as a way of limiting the financial impacts arising from these natural hazards.

Whilst most investigations focus on the physical effects of flooding and aspects such as reduced crop yields, there are other facets that accompany agricultural flooding. In circumstances where crops may not be lost entirely there can be losses associated with not being able to harvest crops at their peak, or having to harvest with poorer quality. With saturated ground and damp crops the susceptibility to fungal diseases as well as other infestations cannot be dismissed and would involve additional treatment such as spraying, leading to increased stress on crops with concomitant decrease in yields. Water logging can also have an adverse effect on soil fertility through increasing the loss of nutrients such as nitrogen. The longer soils remain water logged with low oxygen concentrations the greater the loss of nitrogen from soils.

At the field or farm level measures that can be adopted to reduce flood risk could consider how better management of flood pathways could contribute. The impact of overland flow can be mitigated through good soil management, the slowing down of overland flow through cross-contour interception measures and the maintenance of drainage infrastructure, and, in some cases, harvesting and storage of run-off which could then be used at other times of the year for supplementary irrigation. At the landscape scale, which would require collective action, the aim should be to reduce flood peaks and slow down the flow of water from different parts of the catchment. This could entail temporary storage areas as well as cross-contour structures, as already mentioned. The potential difficulty is that there can be the potential for conflict between farm scale and catchment scale measures. Removing water efficiently and effectively from individual fields or farms could, in aggregate, exacerbate flooding in other parts of a catchment. The other difficulty is in there being a viable mechanism that supports collective action between stakeholders. This has to be supported by a high degree of knowledge and understanding of how the catchment functions and how potential interventions might impact a catchment. Without a well supported, evidence based and coordinated approach, changes would either have little overall effect or exacerbate downstream flooding.

In cases such as the Trinidad example, where downstream agricultural activities are potentially at risk of flooding as a result of upstream actions an approach that is gaining wide acceptance is mechanisms whereby upstream users are remunerated or in other ways rewarded for changes in upstream land management that reduce flood risk. Such schemes are known as Payment for Ecological or Environmental Services (PES). They are most likely to be appropriate where there are clearly definable and measureable benefits for both upstream and downstream stakeholders, noting that it is unlikely that such schemes would have much impact in the case of extreme events described above.

In PES schemes there has to be a buyer and a seller of the services, an identifiable and understood environmental good or service that is to be traded, and a mechanism through which the trade takes place. A more detailed outline of the conditions required for the development of a PES is outlined by Claassen et al. (2005) in which associated issues and objectives together with payment instruments are described. One of the key assumptions is that there is an incentive, realised through the payment mechanism, that encourages land managers such as farmers to undertake land management practices that support the development, conservation or protection of environmental services such as watershed protection. Not only that, but there is an incentive for actively improving the quality of the environmental services. The idea is that through PES the disincentive of receiving no benefit however land in a catchment is managed is countered. One of the best known and longest running example is that of FONAG in Ecuador (see http://www.fonag.org.ec/portal/lang-en/home.html) which has been running since 2000. This provides co-financing of activities, projects and programmes for the rehabilitation, conservation and maintenance of the watersheds that supply water to Quito. In the Caribbean the idea of adopting PES had been explored through a series of papers focusing on watershed protection measures in Grenada, Jamaica, St Lucia, and Trinidad and Tobago (see Geoghegan, 2005; McIntosh and Leotaud, 2007). However, in spite of concluding that there was potential in the region, it has also been found that countries in the region may not be ready or willing to adopt PES (McIntosh and Leotaud, 2007) and that the scope was likely to remain limited.

7. Climate Change

It is almost obligatory when dealing with a subject in which climate plays a major role to mention the possible impacts of climate change. Given that agricultural productivity is affected by climatic parameters such as temperature, precipitation, evaporation, sunlight, wind, seasonality and variability and that flooding could not occur without rainfall, it is appropriate to briefly consider what the potential effects might be. There is a growing body of work that sets out what the climate models have to say about possible future climate change in the Caribbean (e.g., ECLAC, 2010b) and how the changes might vary across the Caribbean.

Studies carried out by the Institute of Meteorology in Cuba and the University of the West Indies (Taylor et al., 2007) have indicated that the mean temperatures across the Caribbean territories have had an upward trend over the last three decades, driven largely by the steady increase in daily minimum temperature values. The studies also showed that the frequency of droughts has increased significantly whilst the frequency of other extreme events in the region seems to be changing, with flooding events and hurricane passage through the region increasing since the mid 1990s (Taylor et al., 2007). Results from regional climate modelling using the United Kingdom Hadley Centre's PRECIS model indicate a likelihood, by 2080, of an annual warming of between 1° and 5° C, depending on the region and scenario. The warming would be greater in the northwest Caribbean territories of Cuba, Jamaica, Hispaniola and Belize than in the eastern Caribbean island chain. Also, there would be greater warming in the summer months than in the cooler and traditionally drier earlier months of the year.

The changes in annual average precipitation suggest a drying across the Caribbean basin. The decreases in rainfall range from 25 to 50% depending on the scenario and part of the Caribbean basin. The southern Caribbean and the islands from St Kitts to Martinique show the largest percentage decreases, relative to the mean precipitation for the climate period 1960 – 1999 baseline by the 2080s. The exceptions to the overall drying trend are the far north: western Cuba and the southern Bahamas, and Costa Rica and Panama – up to 25% wetter under the scenarios. The effect of climate change appears to be to enhance the existing climatic pattern; making the wet and dry zones wetter and drier respectively, during the first 4 – 6 months of the year. However for the months of May to October the entire Caribbean is shown to be up to 25% drier. The changes in average rainfall show a pronounced north-south gradient in rainfall change during the January to April dry season, whilst the summer drying is set to become more severe during the wet season (Taylor et al., 2007). In the case of the Eastern Caribbean the projections are for a substantially drier wet season (July-November) and an even drier dry season (March-April).

Decreases in overall rainfall, especially during the important rainy season, plus an effective lengthening of the dry season period, and higher temperatures and rates of evapotranspiration will not only cause greater ecological stress but also lead to what are likely to be significant reductions in available water resources. The changes in precipitation patterns will also affect run-off characteristics. A predicted increase in the number of no-rainfall days would suggest that when rain does fall it will be more intense, which would result in flashier run-off, increased erosion and potentially greater occurrence of flash flooding. Less overall rainfall and higher rates of run-off would lead to a corresponding decrease in water available for recharging groundwater aquifers as well as affecting surface water flow patterns.

The potential effect of climatic change on crops and crop yields, in the absence of climate -related disasters is uncertain and open to debate. This is due to the fact that complex interaction between aspects such as levels of carbon dioxide, changes in evapotranspiration, and other factors affecting crop production, are far less clear. Also it is not at all clear how such changes might contribute to the risk of flooding. Some have suggested that it may contribute to flooding severity (Betts et al., 2007) but not enough is known and the uncertainty must be acknowledged.

Conclusion

The likelihood of greater climatic variability together with more intense rainfall and run-off events suggest that agriculture will become more vulnerable and prone to the adverse effects of flooding, especially for those areas that are already at risk. The effects of natural disaster s, not just on agriculture are evident and severe: they affect entire economies; lead to loss of production; and put livelihoods at risk. It is all the more imperative, therefore, that farmers and authorities adopt mitigation and adaptation measures that reduce the adverse effects of flooding and increase resilience to enable the agricultural sector to recover. Such measures should be considered as ‘no regret’ measures as they yield benefits even in the absence of increased climate risks and greater variability.

References

Baker, J. (2007). The wet or the dry? Agricultural intensification in the Maya Lowlands. *Studies in Human Ecology and Adaptation*, Volume *3*, 63-90,

Betts, R., Boucher, O., Collins, M., Cox, P., Falloon, P., Gedney, N., Hemming, D., Huntingford, C., Jones, C., Sexton, D. & Webb, M. (2007). Projected increase in continental runoff due to plant responses to increasing carbon dioxide. *Nature*, *448* (7157), 1037 - 1041 k.

Claasen, M., Damon, M., King., Letsoalo, A., Moilwa, N., Ramoelo, A. & Visser, A (2005). The feasibility of developing payments for catchment protection services and improved livelihoods in South Africa. *South African Working Paper Series*, Paper 1. International Institute for Environment and Development, London, UK. pp.114.

Critchley, W. & Bruijnzeel, L. A. (1997). Environmental Impacts of converting moist tropical forests to agriculture and platations. *UNESCO IHP Humid Tropics Program Series* No. *10*, Netherlands.

Deep Ford, J. R., dell'Aquila, C. & Conforti, P. (2007). Agricultural trade policy and food security in the Caribbean : *Structural issues, multilateral negotiations and competitiveness*. Food and Agricultural Organization, Rome, Italy.

Dillehay, T. D., Eling, H. H. & Rossen, J. (2005). Preceramic irrigation canals in the Peruvian Andes. *Proc Natl Acad Sci*, *102*(47), 17241-17244.

Dong, F. & Fuller, F. (2007). Dietary Changes in China 's Cities: Empirical Fact or Urban Legend? Accessed 3 May 2012 from www.card.iastate.edu/iowa_ag_review/winter_07/ article2.aspx

ECLAC. (2003). *Handbook for Estimating the Socio-economic and Environmental Effects of Disasters*, http://www.eclac.cl/id.asp?id=12774

ECLAC. (2006). Guyana : the impact on sustainable livelihoods caused by the December 2005 - February 2006 flooding, LC/CAR/L.088. UN Economic Commission for Latin America and the Caribbean, Sub-regional Headquarters for the Caribbean, Trinidad.

ECLAC. (2010a). Analysis of Extreme Events in the Caribbean 1990 – 2008. LC/CAR/L.254. UN Economic Commission for Latin America and the Caribbean, Sub-regional Headquarters for the Caribbean, Trinidad.

ECLAC. (2010b). Regional Climate Change Modelling, LC/CAR/L.265. UN Economic Commission for Latin America and the Caribbean, Sub-regional Headquarters for the Caribbean, Trinidad.

ECLAC. (2011). Saint Lucia: Macro socio-economic and environmental assessment of the damage and losses caused by Hurricane Tomas: A geo-environmental disaster - Towards resilience, LC/CAR/L.286. UN Economic Commission for Latin America and the Caribbean, Sub-regional Headquarters for the Caribbean, Trinidad.

Falkenmark, M. & Rockström, J. (2004). Balancing Water for Humans and Nature: The New Approach in Ecohydrology. Earthscan, London, UK. pp247.

FAO. (2010). "Climate-Smart" Agriculture: Policies, practices and Financing for Food Security, Adaptation and Mitigation. FAO, Rome, Italy. 41pp.

Geoghegan, T. (2005). Challenges to establishing markets for watershed services : *Learning from country diagnostics*. International Institute for Environment and Development. London, UK.

Gomez-Pompa, A., Luis-Morales, H., Jimenez-Avilla, E, & Jimenez-Avilla, J. (1983). Experiences in traditional hydraulic agriculture. In Kent V. Flannery (Ed.). *Maya subsistence* (327-342). New York: Academic Press.

IFAD. (2011). Rural Poverty Report 2011: *Overview International Fund for Agricultural Development*, Rome, Italy. pp11.

IMF. (2012). Dominica-Request for Disbursement under the Rapid Credit Facility – *Staff Report and Press Release*. IMF Country Report No. 12/47. International Monetary Fund, Washington, D.C.

Jackson, R. (2005). Managing Natural Hazards in Jamaica. Office of Disaster Preparedness and Emergency Management, Kingston, Jamaica accessed on 20 May from http://mona.uwi.edu/cardin/virtual_library/docs/1091/1091.pdf

Lugo, A., Schmidt, R. & Brown, S. (1981). Tropical Forests in the Caribbean. *Ambio*, *10*(6), 318-324.

McIntosh, S. & Leotaud, N. (2007). Fair deals for watershed services in the Caribbean. *Natural Resource Issues* No. *8*. International Institute for Environment and Development. London, UK.

Morris, J., Hess, T. & Posthumus, H. (2010). Agricultures Role in Flood Adaptation and Mitigation: *Policy Issues and Approaches*. Cranfield University, UK. pp35.

O'Connell, E., Ewen, J., O'Donnell, G. & Quinn, P. (2007). Is there a link between agricultural land use management and flooding ? *Hydrology and Earth Systems Science*, *11*, 97-107.

Ramlal, B. & Baban, S. (2008). Developing a GIS based integrated approach to flood management in Trinidad, West Indies, *Journal of Environmental Management*, *88*(4), 1131-1140

Roseboom, J., Cremers, M. & Lauckner, B. (2001). Agricultural R&D in the Caribbean : An Institutional and Statistical Profile. *ISNAR Research Report* No. 19. The Hague: International Service for National Agricultural Research.

Taylor, M. A., Centella, A., Charlery, J., Borrajero, I., Bezanilla, A., Campbell, J., Rivero, R., Stephenson, T. S., Whyte, E. & Watson, R. (2007). *Glimpses of the Future: A Briefing from PRECIS Caribbean Climate Change Project*, Caribbean Community Climate Change Centre, Belmopan, Belize. 24pp.

UNDP. (2006). Human Development Report Beyond scarcity : *Power, poverty and the global water crisis*. United Nations Development Program me, NY, USA.

UNEP/CARICOM. (2004). *Caribbean -Environment Outlook*. United Nations Environment Programme, Kingston, Jamaica. 114pp.

WEF. (2011). *A Vision for Managing Natural Disaster Risk*. World Economic Forum, USA.

In: Flooding and Climate Change
Editors: D. D. Chadee, J. M. Sutherland and J. B. Agard ISBN: 978-1-62948-317-7

Chapter 4

IMPACTS OF FRESHWATER FLOODING IN COASTAL ZONES: SANDY BEACHES, CORAL REEFS AND SEAGRASS BEDS

Judith Gobin*[*]*, Azad Mohammed and John B. Agard
Department of Life Sciences, The University of the West Indies, St. Augustine, Trinidad, West Indies

ABSTRACT

During a major flood in a large river, extremely large volumes of fresh water and sediment can be delivered to near shore areas and the coastal zone. Annual sediment load in the Caribbean is estimated at 10^9 tons per year, which is approximately 12% of the global sediment input from rivers. The riverine inputs bring a combination of freshwater and high sediment loads which, additionally bring an array of potential contaminants from the land run-off. This outflow directly impacts on the marine and coastal zone and associated critical coastal ecosystems (beaches, seagrass beds and coral reefs) in a variety of ways. These flood impacts include (but are not limited to) each and/or a combination of : reduced primary productivity; freshwater bleaching in shallow corals; disease and mortality from high sediment and nutrient loads; increased algal and seaweed blooms; smothering of corals and seagrass; reduced coral and seagrass growth ; negative effects on breeding in some faunal species ; increased levels of pollutants ; increased turbidity; lowered oxygen content; and increased Biological and Chemical oxygen demand. This Chapter deals with some of these negative impacts and concludes by making recommendations for reductions and mitigation measures of the negative impacts.

[*] Corresponding author: Email: Judith.Gobin@sta.uwi.edu.

INTRODUCTION

1.1. How Do We Define the Coastal Zone?

The Coastal Zone can be defined as the set of landward systems which directly affects the marine environment and ecosystems that exist in proximity to land, and usually come under the jurisdiction of that country (Brown et al., 2002). It comprises those areas of continents and islands where the marine and terrestrial environments are in direct contact and includes: the Continental shelf (to a depth of 200m); overlying water mass ; estuaries and deltas; adjacent land including catchments ; offshore islands; cays; and reefs. It can be characterized by both physical (water bodies, currents and tides, climate and weather, rocks and sediments, cliffs, beaches, reefs, floodplains and river mouths) and geomorphological features (prograding coasts, transgressive seas, raised beaches, barrier beaches and sand spits). Within the coastal zone, two basic types of shoreline can be identified: (a) hard rocky areas (e.g., Toco, north-east coast, Trinidad); and (b) depositional shores (e.g., Caroni, Trinidad). The rocky shores are more stable than beaches and generally form cliffs and headlands which are subjected to erosion by the sea. Depositional shores consist primarily of particulate matter which results from erosion and sedimentation accumulation.

Coastal zones include a variety of ecosystems such as mangroves, seagrass beds, coral reefs, mudflats, estuaries, lagoons, littoral woodlands and rocky shores. Each of these represents a dynamic ecosystem with its own diversity of living organisms and non-living components. Often these ecosystems are in close proximity (e.g., Mangal – seagrass – coral reef systems; such as at Buccoo Reef, Tobago) and without obvious boundaries e.g., marine-tidal-terrestrial. In some island states, the land/water interface, where both aquatic and terrestrial systems co-exist, in effect defines the entire island as the coastal zone. In Trinidad and Tobago the coastline is about 552 km long, while the mangrove-coastline is 72 km long. However, the coastal zone covers an area of approximately 36,604 km^2 and includes the wide shelf area around the islands.

1.2. Why Are Coastal Zones Important?

Coastal zones have a variety of important functions; economic services; ecosystem services; primary productivity; recreational services, provision of goods; shoreline protection ; and provision of habitat.

The economical, ecological and social importance of these systems stem primarily from their goods, services, and attributes (Figure 4.1). In the Caribbean, a large proportion of the population resides in coastal areas where industrial activities of strategic national importance, as well as small scale economic activities are also located. The majority of tourist facilities and hotels /guesthouses are also located within coastal areas together with national transportation arteries (coastal roads, bridges etc.). For many countries, economic activities, e.g., extraction of oil and gas, productive fisheries, marine transport and tourism -related activities occur, primarily within the coastal zone. These areas are also impacted by land-based pollution, about 90% of which remains in the coastal zone. Most of the pollutants are transported in bulk by river discharge, effluent discharge, leachate, dumping and atmospheric deposition. This chapter focuses on freshwater flooding in coastal zones, highlighting its impacts on beaches, seagrass beds, coral reefs and general coastal and marine fauna while the topic of mangroves is dealt with in Chapter 5.

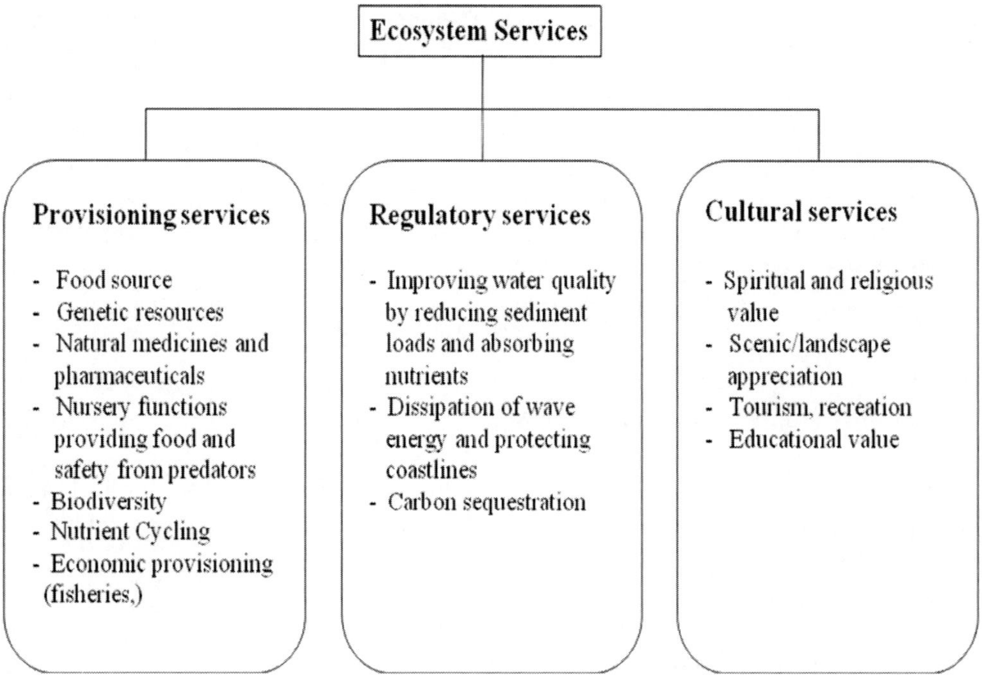

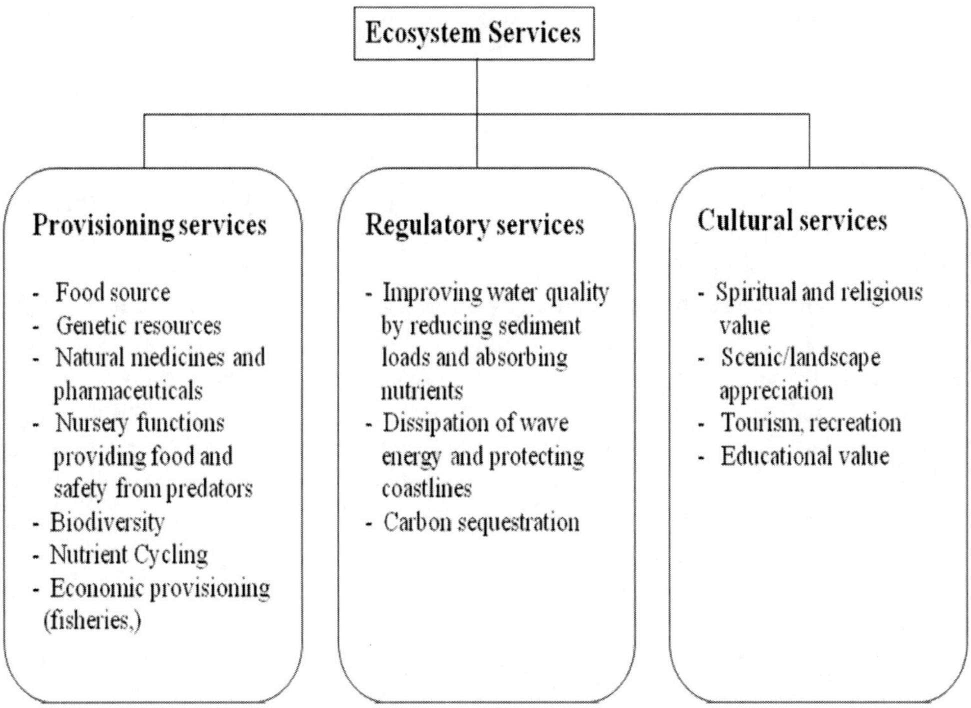

Figure 4.1. Provisioning services of coral reefs.

Primary productivity in coastal and marine waters refers to the amount of carbon fixed by autotrophic organisms through the synthesis of organic matter from inorganic compounds (CO_2 and H_2O), using solar radiation. Primary producers include the dominant phytoplankton, algae, seaweeds, coralline algae, zooxanthellae (symbiotic in corals) and submerged or sub-tidal plants e.g., seagrasses (*Thalassia spp.).* Phytoplankton which constitute the main primary producers in coastal and marine areas consists primarily of single-celled diatoms (Bacillariophycaea) and dinoflagellates (Dinophycaea). Almost all life in the sea depends on the productivity of planktonic organisms which forms the base of the food chain. A number of factors affect primary productivity in the sea:

- Light – the intensity of incident light affects rate of photosynthetic production
- Turbidity - affects the amount of light penetration through the water column
- Nutrients – such as nitrates and phosphates, provide an additional food source for primary producers
- Toxins - toxic compounds can result in increased mortality in marine species
- Freshwater inputs – can alter salinities and result in increased stress in marine species

Increased freshwater inputs is of major concern, as it can change the physical-chemical characteristics of the coastal zone and introduce high levels of land-based contaminants which can significantly affect marine organisms.

Beach characteristics are influenced by wave activity and tidal flows, which affect the movement of sand, grain size distribution and the slope of the beach. Grain size is particularly important, as it affects water retention which is a major factor in determining the distribution of beach organisms. Beaches also serve a vital resource function as some organisms such as *Donax spp.* (chip-chip) and *Emerita spp.* (sea cockroach) are collected for food or bait. Turtle watching (*Dermochelys coriacea* – the leatherback) and bird watching provide important recreational and ecotourism services in many islands. Some turtle species are also harvested as a food source- the green turtle and the loggerhead turtle (*Caretta caretta*), while the hawksbill turtle *(Eretmochelys imbricata)* has a high quality shell used in jewelry and the Olive Ridley (*Lepidochelys olivacea*) possesses high quality skin used in the reptile -leather trade. Some goods (coconut, seagrapes and almonds) are extracted from the strand vegetation (herbs, shrubs). The strand vegetation also supports a rich diversity of insects, lizards, and birds. It also produces leaf litter which adds to the organic pool, which provides valuable nutrients for near shore ecosystems.

Many sandy beaches are under threat from increased storm activity, sand depletion as a result of beach mining activities or increased wave action, and erosion resulting from sea level rise as well as coastal flooding. In the Caribbean, beach sand is used in the construction industry because it is readily available and easily accessible. Beach mining obviously interferes with the beach sediment budget and can leave some beaches depleted of sand for several years until the natural dynamics replace it. The impacts of most of these activities have already been well documented; however few studies have addressed the problem of freshwater flooding.

Seagrasses are rhizomatous marine Angiosperms or true-flowering plants with roots, adapted to live and reproduce while submerged in a saltwater environment. They are unique highly productive ecosystems that capture the sun's energy and convert it into organic matter which serves as the basis for many food web s and are therefore of significant environmental and economic importance. They are among the most productive ecosystems in the world with an average biomass of about 180g C m^{-2} and a net production of about 400g C m^{-2} yr^{-1} or up to 15% of the world's net carbon (Duarte and Chiscano, 1999).

The most common seagrasses found in the Caribbean are *Thalassia testudinum* (turtle grass), *Halodule wrightii* (shoal grass), *Ruppia maritime* (widgeon grass), *Syringodium filiforme* (manatee grass) and three species of the genus *Halophila* (*H. baillonii, H. engelmanni* and *H. decipiens*). These usually form extensive beds in coastal region s with soft marl mud to coarse or shell gravel and coral rubble; with the most luxurious growth occurring in muddy sand in areas with high detritus content. Seagrasses increase the biodiversity (including goods) of coastal waters by providing habitats for a large variety of organisms. They provide grazing and foraging meadows for a variety of fauna including turtles (*Chelonia mydas*), manatees (*Tricheus manatus*), parrotfish (Scaridae), snappers (Lutjanidae), grunts (Scaridae), and commercially important species of queen conchs (*Strombus gigas*), lobsters (*Panulirus argus*) and sea urchin (*Tripneustes esculentus*).

They also provide essential ecosystem services such as:

- Improving water quality, by reducing sediment loads and absorbing nutrients ;
- Stabilizing substrates, by reducing sediment re-suspension and promoting sedimentation ;
- Dissipating wave energy and protecting coastlines;
- Nursery functions providing food and safety (from predators) for juveniles ;
- Acting as an important carbon sinks;
- Nutrient Cycling; and
- Seagrass wrack (leaves detached from the original plant) also plays a vital role as a microhabitat (both in the water and along the shoreline) as well as providing food, and is a source of nutrient cycling for beaches (Bell and Pollard, 1989; Hillman et al., 1989; Duarte, 2002)

Factors that are important for maintaining a healthy seagrass ecosystem include: sediment quality and depth; water quality (temperature, salinity, turbidity); currents; and hydrodynamic processes (water movement); as well as species interactions (epiphytes and grazers).

Coral reefs represent one of the most diverse ecosystems in the world. They consist of a complex collection of interdependent biological communities with distinct zonation and faunal communities. Corals are sessile colonial organisms consisting of many individuals or polyps that secrete a hard calcium carbonate skeleton which is responsible for building up the coral reef over time, and offers protection to the polyps. Coral polyps have a symbiotic (mutually beneficial) relationship with algae (zooxanthellae) which photosynthesise and produce food which is shared with the polyps. In exchange, the coral provides the algae with protection and access to light, which is necessary for photosynthesis. This light requirement generally limits corals to shallow and clear waters. As expected, they do poorly in areas where there is high freshwater and sediment inputs.

Coral reefs can be found distributed throughout the Caribbean region where > 50 species of reef building corals exist. Around Trinidad, reefs are very poorly developed, due to the high turbidity and low salinity of our coastal waters, resulting from large inputs of freshwater from local and South America n rivers. In contrast, well developed coral reefs are found in

Tobago – e.g., the Buccoo Reef which consists of a series of patch reefs forming a barrier around the Bon Accord Lagoon. Coral reefs provide habitats for a large variety of organisms such as molluscs, sea slugs, nudibranchs, oysters, clams, crustaceans (crabs and shrimp), marine worms, echinoderms (star fish and sea urchins), other types of cnidarians (jellyfish and sea anemones), various types of fungi, sea turtles, and various species of fish. All of these organisms depend on the reef ecosystem for food and protection. Because of their high biodiversity, coral reefs are considered to be one of the most productive ecosystems in the world.

Coral reefs have far greater productivity than other marine systems, surpassing 7,000 g C m^2 yr^{-1} (Odum et al., 1959; Helfrich and Townsley, 1965). The net annual primary productivity on coral reefs is estimated to be 2500 g C m^2. It is estimated that the annual global economic value of reefs is about US$375 billion. However, Cesar et al. (2003) estimated the global benefits from coral reefs were about US$29.8 billion annually. Martinez et al. (2007) estimated that in the Caribbean the net economic benefits is between US$3.1-4.6 billion, with dive tourism accounting for about US$2.1 billion, shoreline protection US$0.7 – 2.2 billion and fisheries valued at US$300 million. In Tobago, coral reefs tourism and recreation accounts for US$100-130 million, fisheries US$0.8-1.3 million and shoreline protection US$18-33 million; a substantial proportion of the island's GDP. Reefs also provide a wide range of important ecosystem services (Figure 4. 3) that contributes to the overall development of a country. Globally, reef ecosystems are under threat from a number of factors: e.g., light (depth); salinity ; climate change and increasing sea surface temperatures; ocean acidification; the aquarium trade; blast and cyanide fishing; boating; tourism; land based runoff ; sedimentation and pollutants.

Coastal Flooding

Floods are fundamental events responsible for the transport of freshwater, sediments, solid waste and contaminants from the mainland to the coastal zone. They can be characterized by five (5) major components: *the magnitude of discharge; the velocity of discharge; the duration of the flood; the timing or seasonality*; and *the frequency of disturbance*. Floods can be classified into four (4) types, based on the characteristics of the flow: (1) flash floods which occur over a few hours; (2) single even floods which occur over long durations; (3) multiple event floods which occur frequently (most common in small islands); and (4) seasonal floods.

Individual flood events in combination with human activity such as deforestation, mining and agriculture can result in long-term negative impacts on coastal zone ecosystems resulting from the spread of pollutants, nutrients and sediments (Witt and Siegel, 2000; Schulz, 2001) as well as bacteria and water -borne diseases. Freshwater entering coastal region s through floods can be characterized by low salinity, high turbidity, high sediment load, high nutrients and potentially high contaminant levels. These flood waters often enter the coastal regions as plumes after substantial amounts of rainfall occurs within catchment areas (Figure 4.2).

Figure 4.2. Flood plume from the Caroni River, Trinidad and Tobago (Wayne Rajkumar 2004).

Flood plumes may extend for several kilometers and persist for a few days or several weeks, depending on the frequency and magnitude of the flood. Various types of flood plumes may be observed, depending on the volume of flow, the amount of materials transported in the flood water and the route via which the water reaches the coast:

- Heavy dark brown flood plumes indicative of high very high sediment loads;
- Light brownish flood plumes indicative of light sediment loads; and
- Black, indicative of high organic matter.

These plumes may stay inshore or move offshore depending on the magnitude of the flood, velocity of flow and the prevailing ocean conditions.

Flood water can influence coastal circulation patterns and deliver organic material, pollutants, disease agents and solid waste to the coastal zone, which impacts negatively on marine animals and the food web s supporting them, dramatically shifting ecosystem structure.

2.1. Impacts of Flooding on Coastal Resources (Beaches, Seagrass Beds and Coral Reefs)

Land-based runoff is the main source of fresh water, sedimentation and contamination entering coastal region s. In Trinidad and Tobago, high levels of pollutants in near-shore coastal environments have been related to riverine discharges. Norville (2005) reported that metal input into the Gulf of Paria, occured mainly from urban runoff, industrial and agricultural activity, sewage and domestic wastes and that sites close to rivers typically showed high levels of contaminants. Mohammed et al. (2009, 2011) also reported higher levels of organic contaminants such as organochlorine pesticides, PCBs, dioxins, and Furans at sites close to river outputs in the Beetham. Port-of-Spain Harbour. The mix of nutrients, sediment and pesticides from agricultural run-off, plus unknown amounts of trace metals from flooded areas could have immediate and devastating effects on beaches, seagrasses, coral reefs and marine organisms.

Freshwater outflows also affect primary productivity and its linkages to higher trophic levels, since the major environmental variables (temperature, salinity, light, and nutrient concentrations) which directly influence productivity, are also subject to modification by inputs of flood waters. Flood waters can also cause a wide range of impacts on marine fauna, in particular sessile, slow-moving organisms and benthic species. The observed impacts may include direct responses to exposure to increased sediment loads, toxicants and hyposaline conditions. Pelagic or free swimming organisms are less likely to be impacted directly, as they are able to move away from the flood plumes.

The impacts will be dealt with under the following headings: Physical impacts; Sedimentation impacts including Light reduction; Contaminants and Nutrient increases; and Reduced Salinities. These will be discussed for the main coastal ecosystems as identified above: Beaches, Seagrass beds, Coral reefs and Coastal and Marine organisms.

2.2. Physical Impacts - Beaches

Freshwater flooding can cause significant changes to beach profiles and structure. The location of untrained coastal entrances can vary markedly in response to freshwater floods and coastal storm conditions. At the same time, narrow low sand spits separating estuaries from coastal waters are vulnerable to breakthrough. This is especially so when the catchment has a rapid rainfall response time, which ensures high freshwater discharges with maximum erosion potential. Even where an inlet is stabilized by coastal protection structures, entrance scour due to major floods can have considerable short and long term impacts on the estuary and adjacent shorelines. Flood scour can damage structures built in the active scour zone (Williams, 1981).

Increased water table elevation on the foreshore often occurs following heavy rainfall. At the Stanwell Park beach (New South Wales, Australia), annual rainfall accounted for 12.4% of the variance in the long term high tide position (Gilmour, 1999). The sandy back beach absorbs the rainfall and acts as a slow filter for subsequent discharge of rainwater to the sea through the lower foreshore. Such discharge can lead to decreased sediment shear resistance and then liquidification on the beach face, followed by seaward transport of sand resulting in subsequent shoreline retreat (Syvitski, 2003)

Flood waters transport large amounts of solid waste such as plastics, fibers, fishing gear, metal containers, furniture, styrofoam, vegetation, and appliances etc. into the marine and coastal zone. Some debris (fishing nets, fishing line, plastic bags, strapping bands and 6-pack carriers) can entangle animals resulting in the drowning and death while smaller items (bits of plastic and plastic bottles) may be ingested, causing physical damage and blockage in the digestive tract leading to death. Sharp objects can also cause lacerations and injury.

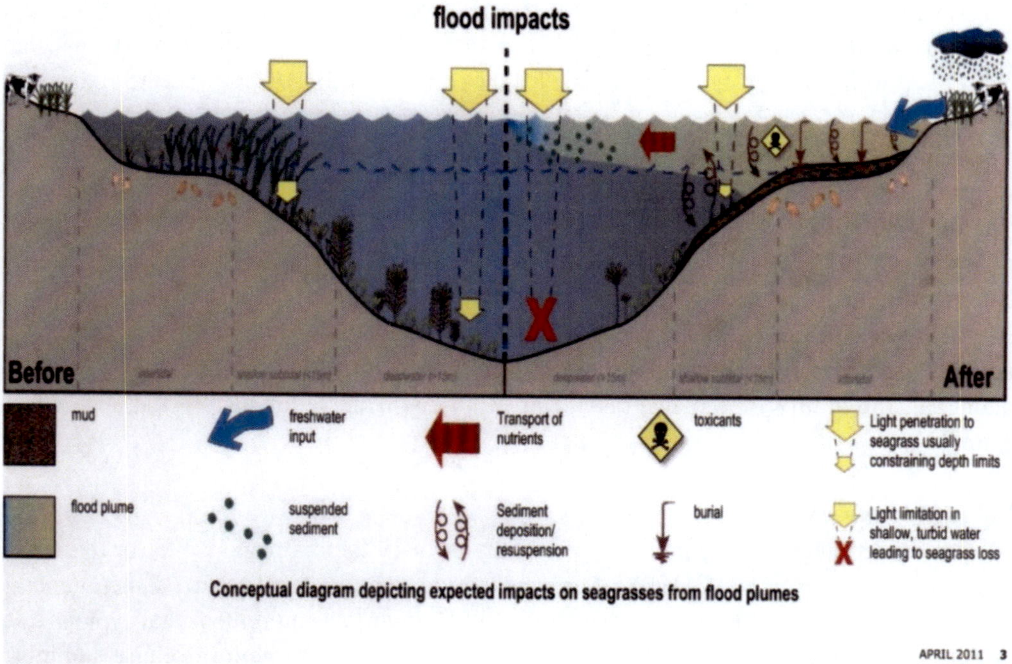

Figure 4.3. Conceptual diagram highlighting the potential impact of flood waters on seagrasses (Mckenzie and Unsworth, 2011).

2.3. Sedimentation Impacts Including Light Reduction

Within the wider Caribbean annual sediment loads are estimated at 1 Gt, or approximately 12% of global sediment input from rivers (UNEP/GPA 2006). The main river systems influencing the Wider Caribbean Region include the Amazon River (Brazil) with a mean sediment load of 1200 million tonnes per year (Luiz et al., 2004), the Magdalena River (Colombia) with a mean sediment load of 143 million tonnes/year (INVEMAR 2006), the Orinoco River (Venezuela) with a mean sediment load of 105 million tonnes/year (MINAMB 2006) and the Mississippi River (USA) with a mean sediment load of 157 million tones /year (Turner et al., 2007). About 80-90% of the particulates entering the coastal zone during flood events consist of fine suspended sediments which can remain suspended in the water column and are often transported over great distances. Due to the small settling velocity of fine particles, they often reach the seafloor only after repackaging and consolidation with other materials.

2.3.1. Seagrasses

Coastal flooding can result in significant alterations of the seagrass ecosystem, negatively impacting on the overall productivity and health of seagrass beds (Figure 4.3).

The impacts of increased sediment loads in the flood waters can be two fold – (1) in shallow near shore environments - the primary effects may be burial of the seagrasses, and (2) in deeper waters, the main effects may be due to reduced light penetration in the submerged habitats. The actual extent of impacts depends on the amount and rate of sediment transport, the persistence of the plume and the prevailing oceanographic conditions. In February 1999, flood waters from the Mary River (Australia) produced a large freshwater plume of suspended sediments which extended 35 km north-west into Harvey Bay (Queensland, Australia) resulting in substantially reduced light conditions for 19 days before returning to pre-flood levels (McKenzie et al., 2000). The authors reported that these conditions caused a dramatic decline in abundance and distribution of seagrasses in the Bay, with a complete disappearance of seagrasses by November 1999. Seagrasses which show di-meristematic forms of leaf replacement (e.g., *Cymodocea nodosa, Thalassia testudinum)* often exhibit increased growth rate s and leaf production rates under moderate sediment burial, although shoot mortality appears to lower overall productivity (Short and Duarte 2001; Mills and Fonseca 2003; Marbà and Duarte 1994; Cabaço and Santos 2007). When the rate of deposition /sedimentation or the volume of sediment deposited per unit area is however greater than the ability of the seagrass to grow through the sediment layer, the plant will eventually die.

2.3.2. Light reduction

Light reduction is considered to be one of the primary factors responsible for the decline of seagrasses (Dennison et al., 1993). Seagrasses require an underwater irradiance generally in excess of 11 % of the surface irradiance (Es) for growth, a requirement that typically sets their depth limit (Duarte, 1991). Increased particulates in the water column cause sunlight to scatter and may completely block the light available for photosynthesis, leading to degradation or death of the seagrasses. Light limitation resulting from sediment laden flood waters may cause a decrease in: above and below ground biomass ; leaf density; leaf length; leaf carbohydrates ; leaf growth rate ; shoot densities; the flowering intensity; and photosynthetic activity if turbid conditions persist for a prolonged period of time (Rollon et al., 2003; Cabello-Pasini et al., 2002; Tomasko and Lapointe 1991; Delgado et al., 1999; Walker et al., 2006; Neverauskas, 1988; Fitzpatrick and Kirkman, 1995; Longstaff and Dennison, 1999; Ruiz and Romero, 2001; Peralta et al., 2002; Kurtz et al., 2003).

Low light conditions can also result in marked physiological changes (e.g., amino acid content, chlorophyll content, and $\delta^{13}C$)-indicative of stress in seagrasses, which are then followed by morphological changes (e.g., biomass, shoot density, canopy height) or die-off (Longstaff and Dennison 1999). Preen et al. (1995) suggested that increased turbidity following cyclonic conditions at Harvey Bay (Queensland, Australia) caused significant losses of *Halophila spinulosa*, *H. ovalis* and *Halodule uninervis.* Similarly, the loss of 95% of a *Z. capricorni* meadow at Sandy Strait (Queensland, Australia) was directly linked with a 2-3 fold increase in turbidity following a flood (Campbell and McKenzie, 2004).

2.3.3. Coral reefs

High sediment inputs accompanying floods can also have significant impacts on coral reefs by adversely affecting their structure and function through alterations of both physical and biological processes. Sediment deposition on coral reefs can generally result in : (a) smothering of marine coral communities and, in severe cases, complete burial and death; increased energy expenditure for surface cleaning by ciliary action; (b) abrasion, decreased calcification, decreased net productivity of corals, and slower rates of reef accretion; (c) lowered light input due to water turbidity resulting in photosynthetic inhibition and shifts in the ecological pattern and depth ranges of growing coral; (d) inhibition of recruitment of corals since the algal-sediment mat reduces the available space for coral larvae attachment or asexual coral expansion; and (e) profound changes in coral population structure s (altered mean colony sizes, altered growth forms, and reduced growth and survival) (Rogers 1990, van Katwijk et al., 1993, West and Van Woesik 2001 Edinger et al., 2000, Anthony and Fabricius 2000, ISRS 2004, Zonta et al., 2005, Fabricius 2005).

In El Rosario Islands (Colombia), increased sediment loads, were responsible for the death of 58% of the coral reef, and between the San Bernardo and Fuerte Islands, live coral decreased by 25% between 1995 and 2000 (CARSEA 2007). Increased sediment not only affects coral populations directly but also indirectly by: increasing macroalgal abundances; bioerosion; increased susceptibility to some diseases; and changes in the abundance of coral predators such as *Acanthaster planci* (Birkeland et al., 1981).

Increased concentrations of suspended sediments can reduce the amount of light reaching corals reefs. At the same time, increased nutrient runoff accelerates phytoplankton production which increases turbidity, reduces light penetration and enhances algal growth, which can cover coral structures (Abal and Dennison, 1996). Although coral reefs can flourish at relatively high levels of particulate matter and silt, they are restricted to the upper 4 m depth in turbid water, while extending to >40 m in clear waters. As light requirements greatly vary between species, few species can tolerate the low light levels at deep depths or at high levels of turbidity. (Cornell and Karlson, 2000). Under conditions of low light, corals settle on upper surfaces where the risk of sediment damage is high, rather than on vertical slopes (Birkeland et al., 1981), which may give rise to compressed zonation.

2.3.4. Coastal and Marine Organisms

Flood borne sediment can be classified into three major groups based on particle size: sand, silt and clay. Coarse particles, like sand and gravel usually settle close to the river outflow, but may be transported along the bottom by the shear velocity of flow. Under turbulent conditions, coarse sediment particles may dislodge sessile organisms and disturb benthic habitats. As silt and clay particles are deposited in the coastal zone they tend to blanket the substrate, thus directly affecting spawning areas and invertebrate habitats by suffocating any eggs or fry. This covering also eliminates interstitial spaces which house organic silt and food on which many other invertebrates feed. Low volumes of sediment may result in organism sublethal and behavioral effects such as: increased activity; stress ; increased emigration rates; loss or reduction of foraging capability; reduced growth and resistance to disease; physical abrasion; clogging of gills; and interference with orientation in homing and migration.

Sedimentation poses one of the greatest challenges to marine organisms, because of its potential for both immediate and long-term effects. Free swimming organisms are often able to avoid these loads although suspended sediments may become attached to gill surfaces clogging them, and eventually causing asphyxiation. Solid particles may also cause physical damage to gill membranes resulting in reduced gas exchange efficiency. High suspended sediment loads also tend to impede feeding abilities because of lower light levels and lower visibility of prey.

Benthic or bottom living communities are directly affected by sedimentation. The increased turbidity which reduces light penetration also reduces primary production of pelagic phytoplankton and benthic algae ; thus reducing key foods for suspension feeders, herbivorous benthic grazers and deposit feeders. Deposited sediments create a fluctuant amorphous layer which may smother the benthic macrofauna which must burrow in order to survive. If the effect of the sedimentation is persistent, then species composition may shift with the opportunistic species becoming the dominant species and resulting in a change in community structure (Mirza and Gray, 1981). For example, Lohrer et al. (2003) reported that *Echinocardium australe* was adversely affected after 3 days (at suspended solids concentrations above 80 mg l^{-1}) with high mortality due to burial. Whereas benthic communities may quickly recover from a single deposition event, a succession of deposition events at shorter intervals than the recovery time can result in cumulative effects that can cause the habitat and the benthic community to change. Goodbody (1961) reported three mass mortality events in Hunts Bay, Jamaica following high rainfall and flood events.

Organisms that live in muddy environments can withstand high concentrations of suspended sediments for short periods of time, but there are still energetic costs (e.g., on the pumping capacity of oysters), which tend to have more chronic rather than acute effects. Those organisms that are found mainly on sandy substrates are often more sensitive to the effects of increased turbidity since they often have to expend greater amounts of energy to clear sediments from their systems. Suspended sediments can also cause oxygen reduction in surface waters from increased biochemical oxygen demand (BOD) associated with increased microbial activities/processes in the water column. This further reduces the amount of oxygen available for marine and coastal organisms.

2.4. Contaminants and Nutrient Increases

2.4.1. Seagrasses

Flood waters transport a range of contaminants (e.g., heavy metals, pesticides pathogens) and nutrients into the coastal zones, and these can have significant impacts on seagrass beds. Heavy metal concentrations decrease fitness in seagrass species and bioaccumulate in plant tissue. Trace metal contamination of seagrasses can impact on associated trophic assemblages through metal transfer and seagrass loss. Heavy metals in water and sediment can both inhibit the growth of seagrass; their metabolic activity; and interfere with vital biochemical pathways, such as photosynthesis. For example, metals such as lead (Pb) and copper (Cu) have been linked to reduced growth rate, leaf size and increased leaf asymmetry in *Halophila ovalis* (Ralph and Burchett, 1998; Ward, 1989 Ambo-Rappe et al., 2011). Generally, when seagrasses are exposed to petrochemicals, sub-lethal quantities are incorporated into the tissue, causing a reduction in tolerance to other stress factors or a decrease in the density of leaves and flowering shoots (Zieman et al., 1984; Dean et al., 1998). These hydrocarbons can coat the leaves and can breakdown the protective wax layers thereby inducing photosynthetic stress.

2.4.2. Coral Reefs

Coral reefs can be impacted by a variety of contaminants including nutrients, pesticides, heavy metals, hydrocarbons and pathogens - all of which significantly alter water quality. This can directly impact the transmission and admission of chemical signals in corals, upsetting the timing of gametogenesis and the synchronisation of the release of gametes (Richmond 1994, Richmond 1997). Some contaminants are: also toxic to corals and larvae ; may reduce their viability and successful settlement; or even have detrimental impacts on photosynthesis. For example, low concentrations of herbicides have been shown to reduce the photosynthetic efficiency of coral symbionts (the dinoflagellates *Symbiodinium* spp.). Higher concentrations of herbicides can also impair the interaction between corals and zooxanthellae, resulting in the expulsion (or loss) of symbionts from the coral host (bleaching) (Owen et. al. 2003).

Coral tissue and their calcium carbonate skeletons can accumulate metal contaminants present in the water column. Several studies have demonstrated that metals substantially accumulate in the zooxanthellae and animal tissue fractions of branching corals (Reichelt-Brushett and McOrist, 2003; Anu et al., 2007; Mitchelmore et al., 2007). However, if corals are exposed to high concentrations of toxic metals, then the zooxanthellae can be expelled resulting in a reduced metal concentration in the corals (Peters et al. (1997); Reichelt-Brushett and McOrist, 2003). This suggests that the expulsion of zooxanthellae may be a mechanism of metal detoxification, since zooxanthellae have been found to accumulate heavy metals to a larger extent and thus be more tolerant than their symbiotic hosts.

The impacts of pesticides in the marine and coastal environments include: changes in reef community structure (such as decreases in live coral cover); bioaccumulation and direct toxicity. Pesticides such as Diuron have been shown to cause photoinhibition, mortality, bleaching, a 5-fold reduction in total lipid content and 6-fold reduction in fecundity in some coral species such as *Acropora tenuis, A. valida,* and a brooding coral, *Pocillopora damicornis* (Cantin et al., 2007). Markey et al. (2007) also suggested that settlement and metamorphosis in *Acropora millepora* was reduced by between 50 and 100% following 18 h exposure to very low concentrations (0.3 to 1.0 μg l–1) of various insecticides and fungicides.

2.4.3. Seagrasses

In pristine environments where nutrient availability may limit biomass production, seagrasses are good competitors, since they have low nutrient requirements and efficiently recycle nutrients. Although some studies have demonstrated a positive relationship between nutrient availability and the growth of tropical seagrass species (e.g., Udy et al., 1999), it is generally believed that anthropogenic increases in nutrients are detrimental to the health of seagrasses and the ecosystems they support (Ralph et al., 2006). However, increased nutrient loads resulting from anthropogenic activities (river runoff and floods in this instance), create an environment where growth is no longer restricted by nutrient limitation. In nutrient enriched systems, fast growing phytoplankton, macroalgae and epiphyte species may outcompete seagrasses for nutrients in the water column, and reduce light availability for photosynthesis (Ralph et al., 2006). Increased nutrient loads also enhance eutrophication, and affect the seagrasses in a number of ways:

- nutrients may stimulate dense growths of phytoplankton which can result in a reduction in the amount of light reaching the seagrasses, eventually leading to mortality ;
- nutrients may cause increased growth in small algae and ephipytes attached to the leaves of seagrasses. If these become too excessive they may result in decreased photosynthetic activity or mechanical breakage of the leaves;
- toxic nutrients may kill invertebrates that feed on the algal epiphytes, causing a sharp increase in the coating of epiphytes, which can weaken, break leaves or cause seagrass mortality; and

- toxic nutrients and sediment anoxia may also contribute to seagrass mortality ; (Cabaço and Santos. 2007).

Juman (2005) reported that during the wet season exceptional runoff carried an increased load of sediment and nutrients into the Bon Accord Lagoon in Tobago where large bare patches of sand in the seagrass beds correlated to high nitrate and phosphate levels. The results also suggested that increased nutrient loads in the lagoon may have enhanced the proliferation of fast-growing algae, phytoplankton epiphytes, and macroalgae which outcompeted the seagrasses.

Eutrophication processes can also lead to an increase in levels of ammonium in sediments and water, which can be toxic to seagrasses. *Zostera marina* for example, becomes necrotic after two weeks exposure to high ammonium concentrations (125 μM) after five weeks, plant size is reduced and they eventually die (van Katwijk et al., 1997).

Excessive nutrients in water can cause massive algal blooms which reduce water clarity. As the nutrients become depleted, the algae begin to die and decompose using up most of the dissolved oxygen which results in anoxic conditions in the water. Seagrasses which are found adjacent to drainage systems are often significantly impacted because of reduced salinities, increased turbidity, and high concentrations of nutrients and chlorophyll *a* (Wear et al., 2006). Approximately 90% of seagrasses disappeared due to a 2-3 fold increase in turbidity and nutrient concentration, and a reduction in light penetration up to 6 months after the flood in Australia (Campbell and McKenzie, 2004).

2.4.4. Coral Reefs

Nutrient (e.g., nitrogen and phosphorous) inputs into the coral ecosystem can result in enhanced eutrophication resulting an algal blooms which can either smother the corals or decrease light penetration. Increased nutrient stress can also result in: (i) local declines in coral biodiversity ; (ii) low, or failure of coral recruitment; (iii) high rates of mortality ; (iv) increases in benthic macroalgae, cyanobacteria and algal blooms; (v) accelerated bioerosion; (vi) loss of reef structure and habitat for reef-dependent organisms; (vii) failure of successful fertilization of gametes; and (viii) changes in reef community structure (Pastorok and Bilyard, 1985; Hallock et al., 1993; Szmant, 2002; Fabricius and McCorry, 2006). In Barbados, increased eutrophication has been the main cause of a shift in the sex ratio of corals *(P. porites)*. On eutrophic reefs there was a skewed sex ratio of 2:1 in favour of males compared to a 1:1 sex ratio on control reefs (Tomascik and Sander 1987). In areas where persistent nutrient inputs occur, such as in reefs influenced by rivers, the reef community may be dominated by hermaphroditic corals, which are better able to survive.

2.4.5. Coastal and Marine Organisms

These toxic substances (heavy metals, nutrients, pesticides, herbicides), can have a number of impacts on biological organisms which include:

- Direct toxicity – where exposure concentrations exceed the tolerance level for individual organisms
- Direct coating and smothering of organisms (e.g., hydrocarbons)

- Ingesting, absorption and bioaccumulation of toxins – low levels may be taken up from the environment and incorporated in tissue or sequestered in different parts of the body
- Ingesting and blockage of digestive system by solid waste
- Low levels of contaminants may induce sublethal responses such as physiological, biochemical or histopathological changes

2.5. Reduced Salinities

Large inputs of freshwater during floods can significantly lower the salinity in the coastal zones. In Trinidad, freshwater discharge from local and South America n rivers following heavy rainfall causes significant seasonal variations in surface water salinity. Wet season salinity may vary between 15-27ppt while dry season salinities vary between 29-35ppt. Following heavy floods, salinities in the south coast may fall to about 10% that of oceanic water while in the Gulf of Paria, salinities may be reduced to 50% that of normal seawater. On the more exposed North Coast, salinity values may drop to 75% that of normal seawater following floods.

2.5.1. Seagrasses

Seagrasses are highly tolerant to short-term exposure to a wide range of salinities (0-45ppt) but tolerance levels may vary between species. Lirman and Cropper (2003) reported marked differences in the tolerance between species exposed to a range of salinities between 5-45ppt, with *S. filiforme* being the most sensitive, then *T. testudinum* while H. *wrightii* and *Z. marina* can tolerate a wide range of salinities. Freshwater inputs may therefore affect density and spread in single species stands, while in multi-species complexes, competitive interactions may be altered resulting in exclusion /replacement of some species based on differential salinity tolerance. *H. wrightii* which appears to be more tolerant to salinity changes is more likely to be dominant in areas prone to frequent salinity changes.

Rapid changes in salinity may not result in immediate effects, but only after prolonged exposure would stress and mortality become evident. The actual extent of the effect depends on the duration of exposure, the extent of salinity reduction, seagrass species and the life stage exposed to salinity reduction. Salinities as low as 13 ppt (approximately 1/3 natural seawater) were shown to have no immediate effect on the growth of *P. australis* (Tyerman et al., 1984). Similarly, adult *A. antarctica* and *P. angustifolia* exposed to salinities as low as 0 ppt, for 72 hours showed no reduction in photosynthetic efficiency (O'Loughlin 2004), however longer exposure, up to 10 days, showed reduced photosynthetic efficiency with plant death occurring within seven weeks. It is therefore possible that seagrass losses due to reduced salinity is most evident very close (within 100 m) to freshwater sources (waste and storm water outfalls and riverine inputs) where salinity levels remain low for long periods of time.

While short-term exposure to lower salinities (during periodic floods) may not be acutely toxic to adults, the same cannot be said for propagules and seeds. Studies have suggested that 72 hours exposure to lowered salinity is enough to inhibit the seedlings of *Amphibolis spp.* and kill the fruit from *Posidonia spp.*. Low salinity also has a greater impact on *A. antarctica* seedlings and *P. angustifolia* fruits which die after seven weeks of exposure to salinities of ~ 1 ppt than on adult plants which are highly tolerant to short-term reductions in salinity.

Anatomical and physiological changes can also be used as indicators of stress in seagrasses exposed to reduced salinities. For example, adult plants of *Z. marina, A. antarctica* and *P. sinuosa,* show a loss of photosynthetic capability, browning of leafage and the loss of leaves following long-term exposure to 0–1 ppt seawater (van Katwijk et al., 1999).

Seagrasses typically have no salt-secreting glands and absorb ions through their leaves. They actively transport sodium, potassium, and chloride ions into channels associated with invaginated plasmalemmas containing numerous mitochondria. Some species also actively exclude salt by an ATP -requiring ion transport mechanism (Jagels, 1973; Jagels, 1983; Jagels and Barnabas, 1989. This exchange of ions allows for the maintenance of stable membrane potentials. However, *h*ypo-osmotic stress can still cause increases in cell volume and turgor, resulting in the loss of ions and organic solutes as well as damage to membranes and organelles, culminating in cell rupture. Fernandez et al. (2006) reported a 50% decline in seagrass meadows (*Cymodocea nodosa*) and the sea urchin (*Paracentrotus lividus*) in the Mediterranean following an exceptionally high rainfall event in which the volume of freshwater that poured into the lagoon corresponded to 36% of its volume.

2.5.2. Corals

Corals are typically found in seawater ranging in salinities from 25 to 42 per cent. Salinity reduction over a short time frame (minutes to hours), or over an extended period of several weeks following flood events, can have substantial impacts on corals and reef ecosystems. Goodbody (1961) reported that heavy rainfall and coastal runoff reduced surface salinity levels in the inshore waters of Jamaica to 5.4 ppt during 1958, and it was 3 months until all affected areas returned to normal levels of salinity. On the Great Barrier Reef plumes of reduced salinity seawater has been found to persist for up to 3 weeks following cyclones (Van Woesik et al., 1995, Devlin et al., 1998).

Salinity reduction can result in mortality, bleaching events (loss of zooxanthellae), excessive mucous release, water gain, cellular pathology and histopathology. These impacts can be exacerbated by other flood-related stressors such as sedimentation, turbidity and increased exposure to ultraviolet radiation. For example, shallow reefs in the Great Barrier Reef Marine Park (Keppel Island region) suffered almost complete mortality due to prolonged salinity stress following the 1991 Fitzroy River floods, and 50 per cent mortality of the fringing reefs around Great Barrier Reef (Snapper Island) occurred due to freshwater flood run-off in 1998. Many studies have also reported that exposure to flood waters cause corals to be discoloured or appear white – a response which is commonly interpreted as bleaching (Goreau 1964, Egana & DiSalvo 1982, Engebretson & Martin 1994, Van Woesik et al., 1995). These bleaching events can occur following decreases in salinity for short periods (i.e., minutes to hours) and slight decreases in salinity for long periods (i.e., days). Such coral bleaching following exposure to low-salinities has been confirmed through laboratory-based studies (Marcus and Thorhaug 1981, Coles 1992, Titlyanov et al., 2000).

Lowered salinities may also cause a reduction in the photosynthetic efficiency of symbiotic dinoflagellates. Generally, morphological changes to hypo-salinity showed: (i) increased tissue swelling, degradation, and loss of zooxanthellae with decreasing salinities,

(ii) paling (bleaching) of the brownish-yellow coloration of the tissue associated with maintenance of a healthy population of zooxanthellae, and (iii) tissue necrosis at the two lowest salinity concentrations (20 and 24 ppt)(Iyer and Barnabas 1993, Tyerman et al., 1984).

2.5.3. Coastal and Marine Organisms

Floodwaters can result in a periodic hypo-osmotic stress in coastal zones which can have significant impacts on survival, functioning, reproduction and development of organisms. Flood waters can change the physicochemical characteristics (salinity, temperature, pH) of sea surface waters, significantly impacting planktonic species and larvae of marine organisms. Many of these may be flushed out of the system or die due to significant reduction in salinity. For example, hypo-saline conditions resulting from heavy rainfall events is reported to induce bleaching, reduced photosynthetic efficiency and altered metabolism of respiratory pathways in anemones (Engebretson and Martin 1994). These metabolic pathologies can, in turn, influence higher-order physiological diseases, such as gamete abnormalities and reduced viabilities, significantly affecting fecundity (Richmond, 1994).

The salinity of coastal waters is also important because of the physical and chemical interaction s that may occur with other stress factors and toxicants. Physical interactions include the effects of salinity on the solubility, uptake, and bioavailability of toxic compounds in aqueous media, while chemical interactions include the modification of the chemical speciation of trace metals (Whitehouse 1984; Nugegoda and Rainbow 1989, Hong and Kester 1985). In addition, salinity affects many responses of organisms (e.g., survival, reproduction, behaviour, and other sublethal effects) to a variety of toxic substances. Reduced salinity can also impact larval development, metamorphosis fertilization and cleavage. *Echinarachnius parma,* for example has a lower limit for successful fertilization and cleavage of about 24psu. At salinities of 20-22 psu, there appears to be a delay in cleavage and less than 10% of the fertilized embryos show cleavage (Allen and Pechenik 2010). At salinities less than 20 psu, no eggs produced fertilization envelopes or cleaved, even 10 h after insemination. Although few studies have explicitly examined the link between salinity and the ability to produce fertilization envelopes, a reduction in fertilization success between 18 and 24 psu does not appear to be unusual for echinoderms. For example, in the asteroid *Asterina pectinifera*, fertilization envelopes do not form at salinities below 18 psu (Kashenko, 2006). Similarly, in the holothuroid *Eupentacta fraudatrix*, 20 psu marked the lower salinity limit for successful fertilization (Kashenko, 2000). For the echinoid *Echinocardium cordatum*, the lowest salinity at which fertilization envelopes were produced was 24 psu (Kashenko, 2007). In the sand dollar *Scapechinus mirabilis*, fertilization envelopes were produced down to salinities of 22 psu, but embryos maintained at these low salinities did not survive to the blastula stage (Bekova et al., 2004).

MITIGATION MEASURES

How can we reduce flooding and reduce or prevent the negative impacts described above? Some urgent actions are listed below.

- The need for Coastal Zone Management Authorities in Caribbean territories which would employ the overarching Integrated Coastal Zone Management goals and objectives, and especially those pertaining to the impacts of flooding.
- The need for full implementation of Caribbean policies and action plans at National levels (e.g., Planning permissions/Physical infrastucture policy) related to flooding and which impact on coastal zone management ; forestry management; land degradation ; waste disposal etc.
- The need for greater harmonisation of the various legislation e.g., greater national conformity with the regional and global treaties and conventions e.g., CBD, UNFCCC etc.
- The need for greater harmonisation/integration of the various national agencies /entities with respect to roles and duties in respect of Coastal Zone Management.
- The need for improved National institutional arrangements required for sharing of information among government Ministries and agencies in order to carry out critical analyses.

CONCLUSION

Caribbean flooding impact data for the coastal zone resources (as above) is lacking hence the text references included a number of non-Caribbean examples. In addition, a major deficiency in many environmental areas in Trinidad and Tobago and the wider Caribbean region is the failure to systematically collect environmental statistics. For example there has not been a comprehensive data collection of flooding information (measurements, locations etc.). Any flooding adaptation measures that are put in place need to be evaluated for their success. These can then be incorporated as lessons learned in the formulation of future mitigative activities.

A scope of statistics and indicators of flooding and its impacts will provide the essential tools to assist in proposing appropriate mitigation measures. For example, the following statistics would be required under Coastal Zone Flooding:

- No. of coastal flooding incidents
- No. of coastal households and persons directly affected
- No. of fishing communities (persons) directly affected and
- No. of livelihood activities (including fishing) affected

The Need for Greater Awareness

Environmental management which includes flood management is a cross cutting task in the Caribbean and Trinidad and Tobago. Although one agency is given the principal responsibility for environmental management, including flood control, related responsibilities are also vested in a few sectoral agencies e.g., ODPM, WRA (see Chapters 1, 2, 3, 6, 9). Coordination is, therefore, key in terms of how information dessimination and public outreach will be facilitated.

Implementation of any of the above (and additional) measures including policy and legislation, requires dedicated technical support. This may be broadly defined to include requirements such as - capacity building and training, co-operation in scientific endeavours e.g., data collection, research, and transfer of technology. Financial Resources pose another challenge to facilitate all of the above and associated responsibilities. The lack of both of these poses a serious challenge to facilitating the proposed activities.

Strengthening of the identified institutional capacity and provision of relevant financial and human resources are therefore the critical shortcomings, to effective functioning and implementation of Caribbean Environmental Management (including Flooding).

REFERENCES

Abal, E. G. & Dennison, W. C. (1996). Seagrass depth range and water quality in southern Moreton Bay, Queensland, Australia. *Marine and Freshwater Research*, *47*, 763–771.

Allen, J. D. & Pechenik, J. A. (2010). Understanding the Effects of Low Salinity on Fertilization Success and Early Development in the Sand Dollar *Echinarachnius parma. Biological Bulletin*, *218*, 189–199.

Ambo-Rappe R., Lajus, D. J. & Schreider M. J. (2011). Heavy metal impact on growth and leaf asymmetry of seagrass, *Halophila ovalis. Journal of Environmental Chemistry and Ecotoxicology*, *3*, 149-159

Anthony, K. R. N. & Fabricius, K. E. (2000). Shifting roles of heterotrophy and autotrophy in coral energetics under varying turbidity. *Journal of Experimental Marine Biology and Ecology*, 252, 221-253.

Anu, G., Kumar, N. C., Jayalakshmi, K. V. & Nair, S. M. (2007). Monitoring of heavy metal partitioning in reef corals of Lakshadweep Archipelago, Indian Ocean. *Environmental Monitoring and Assessment*, *128*, 195–208.

Bekova, N. V., Zhuravel, E. V. & Khristoforova, N. K. (2004). Effects of desalination and the detergent sodium dodecylsulphate on the early development of the sand dollar *Scaphechinus mirabilis. Russian Journal of Marine Biology*, *30*, 175–182.

Bell, J. D. & Pollard, D. A. (1989). Ecology of fish assemblages and fisheries associated with seagrasses. In Larkum, A. W. D., McComb, A. J., Shepherd, S. A. (eds.). *Biology of Seagrasses*. Elsevier Science Publishers B. V., New York, 565-609.

Birkeland, C., Rowley, D. & Randall, R. H. (1981). Coral recruitment patterns at Guan. *Proceedings of the 4th International Coral Reef Congress, 2*, 339-344.

Brown, K., Tompkins, E. L. & Adger, W. N. (2002) MakingWaves: Integrating Coastal Conservation and Development. Earth scan Publications, London. 176p

Cabaço, S. & Santos, R. (2007). Effects of burial and erosion on the seagrass *Zostera noltii. Journal of Experimental Marine Biology and Ecology*, *340*, 204–212.

Cabello-Pasini, A., Lara-Turrent, C. & Zimmerman, R. C. (2002). Effect of storms on photosynthesis, carbohydrate content and survival of eelgrass populations from a coastal lagoon and the adjacent open ocean. *Aquatic Botany*, *74*, 149–164.

Campbell, S. J. & McKenzie, L. J. (2004). Flood related loss and recovery of intertidal seagrass meadows in southern Queensland, Australia. *Estuarine Coast and Shelf Science*, *60*, 477 - 490.

Cantin, N. E., Negri, A. P. & Willis, B. L. (2007). Photoinhibition from chronic herbicide exposure reduces reproductive output of reef-building corals. *Mar Ecol Prog Ser*, *344*, 81–93.

CARSEA. (2007). Caribbean Seas Ecosystem Assessment (CARSEA). A sub-global component of the Millennium Ecosystem Assessment (MA), J. Agard, A. Cropper, K. Garcia eds. *Caribbean Marine Studies*, Special Edition, 2007. 104p.

Cesar, H., Burke, L. & Pet-Soede, L. (2003). The economics of worldwide coral reef degradation. Cesar Environmental Economics Consulting. Arnhem, Netherlands.

Coles, S. L. (1992). Experimental comparison of salinity tolerances of reef corals from the Arabian Gulf and Hawaii : evidence for hyperhaline adaptation. Proc 7th Int Coral Reef Symp, *Guam*, *1*, 227–234.

Cornell, H. V. & Karlson, R. H. (2000). Coral species richness : ecological versus biogeographical influences. *Coral Reefs*, *19*, 37–49.

Devlin, M., Taylor, J. & Brodie, J. (1998). Flood plumes, extent, concentration and composition. *GBRMPA Reef Research*, *8*, 1–9.

Dean, T. A., Stekoll, M. S., Jewett, S. C., Smith, R. O. & Hose, J. E. (1998). Eelgrass (*Zostera marina* L.) in Prince William Sound, Alaska : Effects of the *Exxon Valdez* oil spill. *Marine Pollution Bulletin*, *36*, 201-210.

Delgado, O., Ruiz, J., Pérez, M., Romero, J. & Ballesteros, E. (1999). Effects of fish farming on seagrass (*Posidonia oceanica*) in a Mediterranean bay: seagrass decline after organic loading cessation. *Oceanologica Acta*, *22*, 109-117.

Dennison, W. J., Orth, R. J., Moore, K. A., Stevenson, J. C., Carter, V., Kollar, S., Bergstrom, P. W. & Batiuk, A. (1993). Assessing water quality with submerged aquatic vegetation. *BioScience*, *43*, 86–94.

Duarte, C. M. (1991). Seagrass depth limits. *Aquatic Botany*, *40*, 363– 377.

Duarte, C. M. & Chiscano, C. L. (1999). Seagrass biomass and production: a reassessment. *Aquatic Botany*, *65*, 159-174.

Edinger, E. N., Limmon, G. V., Jompa, J., Widjatmoko, W., Heikoop, J. M. & Risk, M. J. (2000). Normal coral growth rate s on dying reefs: Are coral growth rates good indicators of reef health ? *Marine Pollution Bulletin*, *5*, 404-425.

Egana A. C. & DiSalvo L. H. (1982). Mass expulsion of zooxanthellae by Easter Island corals. *Pacific Science*, *36*, 61–63.

Engebretson, H. & Martjn, K. L. M. (1994). Effects of Decreased Salinity on Expulsion of Zooxanthellae in the Symbiotic Sea Anemone *Anthopleura elegantissima Pacific Science*, *48*, 446-457.

Fabricius, K. (2005). Effects of terrestrial runoff on the ecology and coral reefs : review and synthesis. *Marine Pollution Bulletin*, *50*, 125e146.

Fabricius, K. E. & McCorry, D. (2006). Changes in octocoral communities and benthic cover along a water quality gradient in the reefs of Hong Kong. *Marine Pollution Bulletin*, *52*, 22–33.

Fernandez, C., Pasqualini, V., Boudouresque, C., Johnson, M., Ferrat, L., Caltagirone, A. & Mouillot, D. (2006). Effect of an exceptional rainfall event on the sea urchin (Paracentrotus lividus) stock and seagrass distribution in a Mediterranean coastal lagoon. *Estuarine*, *Coastal and Shelf Science*, *68*, 259-270.

Fitzpatrick, J. & Kirkman, H. (1995). Effects of prolonged shading stress on growth and survival of seagrass *Posidonia australis* in Jervis Bay, New South Wales, Australia. *Marine Ecology Progress Series*, *127*, 279-289.

Gilmour, J. (1999). Experimental investigation into the effects of suspended sediment on fertilisation, larval survival and settlement in a scleractinian coral. *Marine Biology*, *135*, 451-462.

Goodbody, I. (1961). Mass mortality of a marine fauna following tropical rains. *Ecology*, *42*, 150-155.

Goreau, T. F. (1964). Mass expulsion of zooxanthellae from Jamaican reef communities after Hurricane Flora. *Science*, *145*, 383–386.

Hallock, P., Muller-Karger, F. E. & Halas, J. C. (1993). Coral-reef decline. *National Geographic Research and Exploration*, *9*, 358–378.

Helfrich, P. & Townsley, S. J. (1965). Influence of the sea, pp. 39-56. In: F.R. Fosberg (ed.), *Man's place in the island ecosystem. 10th Pacific Science Congress* (1961). Bishop Mus. Press, Honolulu. 264 pp.

Hillman, K., Walker, D. I., Larkum, A. W. D. & McComb, A. J. (1989). Productivity and nutrient limitation. In: Larkum, A. W. D., McComb, A. J., Shepherd, S. A. (eds.). *Biology of seagrasses. A treatise of the biology of seagrasses with special reference to the Australian region*. Elsevier, Amsterdam. 635-685.

Hong, H. & Kester, D. R. (1985). Chemical forms of iron in the Connecticut River estuary. *Estuarine Coastal Shelf Science*, *21*, 449-459.

INVEMAR. (2006). Inventario y Caracterización de Fuentes Terrestres Fijas de Contaminación Sobre Las Áreas Marinas y Costeras Del Caribe Colombiano Irlandi, E, Orlando, B., Maciá, S., Biber, P., Jones, T., Kaufman, L., Lirman, D., Patterson, E.T. (2002). The influence of freshwater runoff on biomass, morphometrics, and production of *Thalassia testudinum Aquatic Botany*, *72*, 67–78.

ISRS. (2004). The effects of terrestrial runoff of sediments, nutrients and other pollutants on coral reefs. Briefing Paper 3, International Society for Reef Studies, pp: 18.

Iyer, V. & Barnabas, A. D. (1993). Effects of varying salinity on leaves of *Zostera capensis* Setchell. I. Ultrastructrual changes. *Aquatic Botany*, *46*, 141–153.

Jagels, R. (1973). Studies of a marine grass. *Thalassia testudinum*. Part I. Ultrastructure of the osmoregulatory leaf cells. *American Journal of Botany*, *60*, 1003–1009.

Jagels, R. (1983). Further evidence for osmoregulation in epidermal leaf cells of seagrasses. *American Journal of Botany*, *70*, 327–333.

Jagles, R. & Barnabas, A. (1989). Variation in leaf ultra-structure of *Ruppia maritime* L. along a salinity gradient. *Aquatic Botany*, *33*, 207-221.

Juman R. A. (2005). The structure and productivity of the *Thalassia testudinum* community in Bon Accord Lagoon, Tobago *International Journal of Tropical biology*, *53* (Suppl. 1), 219-227,

Kashenko, S. D. (2000). Combined effect of temperature and salinity on the development of the holothurian *Eupentacta fraudatrix. Russian Journal of Marine Biology*, *26*, 188–193.

Kashenko, S. D. (2006). The combined effect of temperature and salinity on development of the sea star *Asterina pectinifera. Russian Journal of Marine Biology*, *32*, 37–44.

Kashenko, S. D. (2007). Adaptive responses of embryos and larvae of the heart-shaped sea urchin *Echinocardium cordatum* to temperature and salinity changes. *Russian Journal Marine Biology*, *33*, 381–390.

Kurtz, J. C., Yates, D. F., Macauley, J. M., Quarles, R. L, Genthner, F. J., Chancy, C. A. & Devereux, R. (2003). Effects of light reduction on growth of the submerged macrophyte *Vallisneria americana* and the community of root -associated heterotrophic bacteria. *Journal of Experimental Marine Biology and Ecology, 291*, 199-218.

Lirman, D. & Cropper, W. P. Jr. (2003). The influence of salinity on seagrass growth survivorship, and distribution within Biscayne Bay, Florida: Field, experimental, and modelling studies. *Estuaries, 26*, 131-141.

Lohrer, A. M., Hewitt, J., Thrush, S. F., Lundquist, C. J., Nicholls, P. E. & Leifting, R. (2003). Impacts of terrigenous material deposition on subtidal benthic communities. Prepared for Auckland Regional Council. NIWA Client Report HAM2003-055.

Longstaff, B. J. & Dennison, W. C. (1999). Seagrass survival during pulsed turbidity events: the effects of light deprivation on the seagrasses *Halodule pinifolia* and *Halophila ovalis*. *Aquatic Botany, 65*, 105-121.

Martinelli, L. A., Victoria, R. L., Devol, A. H., Richey, J. E. & Forsberg, B. R. (2004). *Suspended sediment load in the Amazon basin: An overview, GeoJournal, 19*, 381-389. Taken 2008-06-10 from http://www.springerlink.com/content/w5q1471116522231

Marcus, J. & Thorhaug, A. (1981). Pacific versus Atlantic responses of the subtropical hermatypic coral *Porites* spp. To temperature and salinity effects. *Proceedings of the 4th International Coral Reef Symposium, Quezon City, 2*, 15–20.

Marbà, N. & Duarte, C. M. (1994). Growth response of the seagrass *Cymodocea nodosa* to experimental burial and erosion. *Marine Ecological Progress Series, 107*, 307–311.

Markey, K. L., Baird, A. H., Humphrey, C. & Negri, A. P. (2007). Insecticides and a fungicide affect multiple coral life stages. *Marine Ecology Progress Series, 330*, 127–137.

Martínez, M. L., Intralawan, A., Vasquez, G., et al. (2007). The coasts of our world: ecological, economic and social importance. *Ecological Economics, 63*, 254–272.

McKenzie, L. J., Roder, C. A., Roelofs, A. J. & Lee Long, W. J. (2000). Post-flood monitoring of seagrasses in Hervey Bay and the Great Sandy Strait 1999. Implications for dugong, turtle and fisheries management. Cairns, Department of Primary Industries, DPI Information Series QI00059.

Mckenzie, L. J. & Unsworth, R. (2011). Surviving the flood: How long can seagrass hold their breath. McKenzie, L.J., Yoshida, R.L., Unsworth, R. (Eds) *Seagrass-Watch News*. Issue 43 April 2011. Seagrass-watch HQ 28pp.

Mills, K. E. & Fonseca, M. S. (2003). Mortality and productivity of eelgrass *Zostera marina* under conditions of experimental burial with two sediment types. *Marine Ecological Progress Series, 255*, 127 – 134.

MINAMB. (2006). Cargas Contaminantes provenientes de las Fuentes Terrestres de Contaminación Marina, Venezuela, Informe Final, Ministerio del Ambiente.

Mirza, F. B. & Gray, J. S. (1981). The fauna of benthic sediment from the organically enriched Olsofjord, Norway. *Journal of Experimental Marine Biology & Ecology, 54*, 181–207.

Mitchelmore, C. L., Verde, E. A. & Weis, V. M. (2007). Uptake and partitioning of copper and cadmium in the coral Pocillopora damicornis. *Aquatic Toxicology, 85*, 48–56.

Mohammed, A., Peterman, P., Echols, K., Feltz, K., Tegerdine, G., Manoo, A., Maraj, D., Agard, J. & Orazio, C. (2011). Polychlorinated biphenyls (PCBs) and organochlorine pesticides (OCPs) in harbor sediments from Sea Lots, Port-of-Spain, Trinidad and Tobago. *Marine Pollution Bulletin*, *62*, 1324–1332.

Mohammed, A., Orazio, C., Peterman, P., Echols, K., Feltz, K., Manoo, A., Maraj, D. & Agard, J. (2009). Polychlorinated Dibenzo-p-dioxin (PCDDs) and Polychlorinated Dibenzofurans (PCDFs) in Harbor Sediments from Sea Lots, Port-of-Spain, Trinidad and Tobago. *Marine Pollution Bulletin*, *58*, 928-946

Neverauskas, V. P. (1988). Response of a *Posidonia* community to prolonged reduction in light. *Aquatic Botany*, *31*, 361-366.

Norville, W. (2005). Spatial distribution of heavy metals in sediments from Gulf of Paria, Trinidad. *Revista de Biologia Tropical*, *53*(Suppl. 1), 33-40.

Nugegoda, D. & Rainbow, P. S. (1989). Salinity, osmolarity, and zinc uptake in *Palaemon elegans* (Crustacea: Decapoda). *Marine Ecology Progress series*, *55*, 149-157.

Odum, H. T., Burkholder, P. R. & Rivero, J. (1959). Measurements of productivity of turtle grass flats, reefs, and the Bahia Fosforescente of southern Puerto Rico. *Publications of the Institute of Marine science*, *6*, 159-171.

O'Loughlin, E. (2004). Stress response of *Amphibolis antarctica* and *Posidonia sinuosa* to freshwater inputs. Honours thesis, Department of Environmental Biology, The University of Adelaide. 72 pp.

Owen, R., Knap, A., Ostrander, N. & Carbery, K. (2003). Comparative acute toxicity of herbicides to photosynthesis of coral zooxanthellae. *Bulletin of Environmental Contamination and Toxicology*, *70*, 541–548.

Pastorok R. A. & Bilyard G. R. (1985). Effects of sewage pollution on coral reef communities. *Marine ecology Progress Series*, *21*, 175-189.

Peralta, G., Pérez-Lloréns, J. L., Hernández, I. & Vergara, J. J. (2002). Effects of light availability on growth, architecture and nutrient content of seagrass *Zostera noltii* Hornem. *Journal of Experimental Marine Biology and Ecology*, *269*, 9-26.

Preen, A. R., Lee Long, W. J. & Coles, R. G. (1995). Flood and cyclone related loss, and partial recovery, of more than 1000 km2 of seagrass in Hervey Bay, Queensland, Australia. *Aquatic Botany*, *52*, 3-17.

Ralph, P. J. & Burchett, M. D. (1998). Photosynthetic response of *Halophila ovalis* to heavy metal stress. *Environmental Pollution*, *103*, 91-101.

Ralph, P. J., Tomasko, D., Moore, K., Seddon, S. & Macinnis-Ng, C. M. O. (2006). Human impacts on seagrasses : Eutrophication, sedimentation, and contamination. *In* : Larkum, A.W.D., Orth, R.J. and Duarte, C.M. (Eds), S*eagrasses : Biology, Ecology and Conservation*, 567-593. Springer, Dordrecht, Netherlands.

Reichelt-Brushett, A. J. & McOrist, G. (2003). Trace metals in the living and nonliving components of scleractinian corals. *Marine Pollution Bulletin*, *46*, 1573–1582.

Richmond, R. H. (1994). Effects of coastal runoff on coral reproduction. In Ginsburg RN (ed) *Proceedings of the Colloquium on Global Aspects of Coral Reefs*. Univ. of Miami, 360-364.

Richmond, R. H. (1997). Reproduction and recruitment in corals: Critical links in the persistence of reefs. In: Birkeland C. (ed). Life and death of coral reefs. Chapman and Hall, New York, USA, 175-196.

Rogers, C. S. (1990) Responses of coral reefs and reef organisms to sedimentation. *Marine Ecology Progress Series*, *62*, 185-202.

Rollon, R. N., van Steveninck, E. D. & van Vierssen, W. (2003). Spatio-temporal variation in sexual reproduction of the tropical seagrass *Enhalus acoroides* (L.f.) Royle in Cape Bolinao, NW Philippines. *Aquatic Botany*, *76*, 339-354.

Ruiz, J. M. & Romero, J. (2001). Effects of *in situ* experimental shading on the Mediterranean seagrass *Posidonia oceanica. Marine Ecology Progress Series*, *215*, 107-120.

Schulz, R. (2001). Rainfall-induced sediment and pesticide input from orchards into the Lourens River, Western Cape, South Africa: importance of a single event. *Water Research*, *35*, 1869-1876.

Short, F. T. & Duarte, C. M. (2001). Methods for the measurement of seagrass growth and production. In *Global seagrass research methods*, ed. F. T. Short and R. G. Coles, 155-182. Amsterdam: Elsevier Science.

Syvitski, J. P. M. (2003). Supply and flux of sediment along hydrological pathways: research for the 21st century. *Global and Planetary Change*, *39*, 1 – 11.

Szmant, A. M. (2002). Nutrient enrichment on coral reefs : is it a major cause of coral reef decline? *Estuaries*, *25*, 743-766.

Titlyanov, E. A., Tsukahara, J., Titlyanova, T. V., Leletkin, V. A., Van Woesik, R. & Yamazato, K. (2000). Zooxanthellae population density and physiological state of the coral *Stylophora pistillata* during starvation and osmotic shock. *Symbiosis*, *28*, 303–322.

Tomasko, D. A. & Lapointe, B. E. (1991). Productivity and biomass of *Thalassia testudinum* as related to water column nutrient availability and epiphyte levels: field observations and experimental studies. *Marine Ecology Progress Series*, *75*, 9-17.

Tomascik, T. & Sander, F. (1987). Effects of eutrophication on reef-building corals. 3. Reproduction of the reef-building coral *Porites porites*. *Marine Biology*, *94*, 77-94.

Turner, R. E., Rabalais, N. N., Alexander, R. B., McIsaac, G. & Howarth, R. W. (2007). *Characterization of Nutrients, Organic Carbon, and Sediment Loads and Concentrations from the Mississippi River into Northern Gulf of Mexico, Estuaries and Coasts*, Vol. *30*, No. 5, 773-790

Tyerman, S. D., Hatcher, A. I., West, R. J. & Larkum, A. W. D. (1984). *Posidonia australis* growing in altered salinities: leaf growth, regulation of turgor and the development of osmotic gradients. *Australian Journal of Plant Physiology*, *11*, 35-47.

Udy, J. W., Dennison, W. C., Lee Long, W. J. & McKenzie, L. J. (1999). Response of seagrass to nutrients in the Great Barrier Reef, Australia. *Marine Ecology Progress Series*, *185*, 257- 271.

UNEP/GPA. (2006), The State of the Marine Environment: A regional assessment, The Hague.

van Katwijk, M. M., Meier, N. F., van Loon, R., van Howe, E. M., Giesen, W. B. J. T. & van der Velde, G. (1993). Sabaki River sediment loading and coral stress : correlation between sediments and condition of the Malindi-Watamu reefs in Kenya (Indian Ocean). *Marine Biology*, *117*, 675-683

van Katwijk, M. M., Vergeer, L. H. T., Schmitz, G. H. W. & Roelois, J. G. M. (1997). Ammonium toxicity in eelgrass *Zostera marina. Marine Ecology Progress Series*, *157*, 159-173.

van Katwijk, M. M., Schmitz, G. H. W., Gassling, A. P. & van Avesaagth, P. A. (1999). Effects of salinity and nutrient load and their interaction on *Zostera marina. Marine Ecology Progress Series*, *190*, 155-165.

Van Woesik, R., De Vantier, L. M. & Glazebrook, J. S. (1995). Effects of Cyclone 'Joy' on nearshore coral communities of the Great Barrier Reef. *Marine Ecology Progress Series, 128*, 261–270.

Walker, D. I., Kendrick, G. A. & McComb, A. J. (2006). Decline and recovery of seagrass ecosystems—the dynamics of change. Pages 551–565 in Larkum AWD, Orth RJ, Duarte CM, eds. Seagrasses: Biology, Ecology and Conservation. Dordrecht (The Netherlands): Springer.

Ward, T. J. (1989). The accumulation and effects of metals in seagrass habitats. In: Larkum AWD, McComb AJ, Shepherd SA (eds.) *Biology of seagrasses: A treatise on the biology of seagrasses with special reference to the Australian region.* Elsevier, New York, 797-820.

West, K. & Van Woesik, R. (2001) Spatial and temporal variance of river discharge on Okinawa (Japan): Inferring the temporal impact on adjacent coral reefs. *Marine Pollution Bulletin, 42*, 864-872.

Williams, G. (1981). "Investigation and Management of Estuarine Lake Inlets". First Australian Conference on Hydraulics in Civil Engineering, Sydney, October 1981. I.E.Aust.

Wear, R. J., Eaton, A., Tanner, J. E. & Murray-Jones, S. (2006) The impact of drain discharges on seagrass beds in the South East of South Australia. Final Report Prepared for the South East Natural Resource Consultative Committee and the South East Catchment Water Management Board. South Australian Research and Development Institute (Aquatic Sciences) and the Department of Environment and Heritage, Coast Protection Branch, Adelaide. RD04/0229-3.

Witt, G. & Siegel, H. (2000). The consequences of the Oder flood in 1997 on the distribution of polycyclic aromatic hydrocarbon s (PAHs) in the Oder River estuary. *Marine Pollution Bulletin, 40*, 1124–1131.

Whitehouse, B. G. (1984). The effects of temperature and salinity on the aqueous solubility of polynuclear aromatic hydrocarbons. *Marine Chemistry, 14*, 19-332.

Zieman, J. C., Orth, R., Phillips, R. C., Thayer, G. & Thorhaug, A. (1984). The effects of oil on seagrass ecosystems. In: Cairns, J. and Buikema, A.L., (Eds), *Restoration of Habitats Impacted by Oil Spills.* 37-64, Butterworth, Boston.

Zonta, R, Collavini, F., Zaggia, L. & Zuliani, A. (2005), The effect of floods on the transport of suspended sediments and contaminants: A case study from the estuary of the Dese River (Venice Lagoon, Italy) *Environment International, 31*, 948 – 958.

In: Flooding and Climate Change
Editors: D. D. Chadee, J. M. Sutherland and J. B. Agard
ISBN: 978-1-62948-317-7

Chapter 5

EFFECTS OF FRESHWATER FLOODING ON MANGROVES IN THE CARIBBEAN REGION

Azad Mohammed**, Aidan D. Farrell, Judith Gobin and John B. Agard
Department of Life Sciences, The University of the West Indies,
St. Augustine, Trinidad, West Indies

ABSTRACT

Mangrove ecosystems are widely distributed in the Caribbean Region, covering *c.* 2.4 million hectares. They provide numerous ecosystem services, notably they act as nursery grounds for many marine and terrestrial animals, prevent coastal erosion, and act as carbon sinks. Mangrove trees have developed a range of morphological and physiological adaptations that allow them to grow in harsh saline, anoxic, usually acidic, and waterlogged environments. While periodic deposits of freshwater sediments provide nutrients for mangrove systems, major flooding can have serious, detrimental, and long-lasting effects. Here we discuss the complex effects that large pulses of freshwater can have on salinity, soil redox potentials, and nutrient availability, as well as the impact of these abiotic factors on the biodiversity and productivity of mangrove systems.

1. INTRODUCTION

1.1. What are Mangroves?

Mangrove ecosystems are dominated by flora and fauna that have developed unique adaptations to cope with frequent saltwater inundation. Mangrove trees are highly specalised plants that are adapted for growing in saline environments ranging from brackish water, through pure seawater (30 ‰ – 35 ‰), to hypersaline conditions (> 60 ‰). They are common in tropical and subtropical regions between latitudes 25 °S and 25 °N, where water

* Corresponding author: Email: Azad.mohammed@sta.uwi.edu.

temperatures are often greater than 24 °C in the warmest months. Mangroves thrive best in areas where the average monthly temperature exceeds 20 °C, seasonal ranges do not exceed 5 °C and annual rainfall exceeds 1250 mm.

Mangrove ecosystems are characterized by soft anoxic sediments that develop due to constant inundation and sedimentation. However, mangrove trees have developed morphological and physiological adaptations that allow them to survive under these conditions. The main adaptations include: aerial roots, aerenchyma, lenticels, salt exclusion, salt excretion, and a range of other mechanism that confirm tolerance to high salinity while maintaining a positive water and carbon balance. In addition, some species are viviparous, producing seedlings that germinate while attached to the parent plant and can establish themselves directly in the sediment at low tide or be dispersed by tidal flows.

Aerial roots provide structural support in the soft sediments and reduce the impact of flooding by enabling the plant to transfer gases from above to below ground and *vice versa*. Many mangrove trees have extensive prop roots which descend from the trunk and branches (e.g., *Rhizhopora spp.;* Figure 5.1). On the surface of these prop roots are numerous lenticels, which allow oxygen to diffuse from the air into gas-filled aerenchyma channels through which it can be transported to belowground tissues. The lenticels are highly hydrophobic and prevent water penetration into the aerenchyma system during high tide. Some species have an extensive system of shallow, underground 'cable roots' that radiate out from the central trunk. *Laguncularia* and *Avicennia spp.* contain special vertical roots, called pneumatophores (Figure 5.2), which project above the soil (to a height of 20-30 cm) permitting oxygen to reach submerged roots. The density, size and number of pneumatophores vary between trees and the aerenchyma may occupy as much as 70 % of the total root volume.

Figure 5.1. Prop roots of the red mangrove (*Rhizophora mangle*) in Trinidad, West Indies.

Figure 5.2. Pneumatophores roots of the black mangrove *(Avicennia germinans)* in Trinidad, West Indies.

Mangroves have also evolved mechanisms to cope with high salinities: salt exclusion, salt excretion, and tolerance of high salt concentration s within tissues. *Rhizophora, Bruguiera* and *Ceriops spp.* have roots that function as ultrafilters, excluding salt while extracting water from soils (Rutzler and Feller, 1996; Kathiresan and Bingham, 2001; Liang et al., 2008). Leaves can store salt in cell vacuoles or use salt-secreting glands that cause salt crystals to form on the leaf surfaces (e.g., *Avicennia* and *Laguncularia spp.*). This salt is subsequently lost when the leaves are abscised (Drennan and Pammenter, 1982; Kathiresan and Bingham, 2001; Liang et al., 2008).

2. Distribution of Mangroves

Mangrove forests cover about 181,000 km^2 (18,100,000 ha) of the earth's surface (Spalding et al., 1997), or about 0.1 % of the total land area. Within the Caribbean, mangroves can be found along most coastlines and cover an area of about 7820.63 km^2 or about 780,236 ha (FAO, 2004). The largest mangrove forests are found in South and Central America, Cuba and Bahamas in areas characterized by low-relief coastal plains that receive an ample supply of freshwater (Table 5.1). The islands of the Eastern Caribbean which typically have low freshwater runoff, steep shore lines and large coastal region s exposed to high tidal and wave activity (e.g., Barbados, Grenada, St Lucia, St Vincent and the Grenadines) generally have smaller mangrove forests (Table 5.1). Other factors which may influence the distribution of mangroves include: temperature, nutrient input, freshwater inflows, and salt tolerance.

Globally, approximately 35 species of true mangroves and 60 species of mangrove-associated plants have been classified. Within the Caribbean, seven mangrove species have been reported: *Rhizopora mangle*, *R. harrisonii*, *R. racemosa*, *Avicennia germinans*, *A. Schaueriana*, *Laguncularia racemosa*, and *Conocarpus erectus* (Bacon, 1993). Of the true mangrove species occurring in the wider Caribbean, red (*Rhizophora mangle*), black (*Avicennia germinans*), and white (*Laguncularia racemosa*) appear to be the most dominant.

Table 5.1. Recent estimates of mangrove cover in the Caribbean region

Country	Year of report	Area coverage (ha)	Source
Anguilla	2005	90	FAO, 2004
Antigua Barbuda	2002	1,200	CCS, 2002
Aruba	2000	420	FAO, 2004
Bahamas	2002	142,000	CCS, 2002
Barbados	2004	4	FAO, 2004
Bermuda	2001	18	Anderson et al., 2001
British Virgin Islands	2001	587	FAO, 2004
Cayman islands	1998	7,830	FAO, 2004
Belize	2010	74,684	Cherrington et al., 2010
Costa rica	2000	41,840	FAO, 2004
Cuba	2009	556,900	Ecoearth
Dominica	2002	10	CCS, 2002
Dominican Republic	2009	32,500	Ecoearth
El Salvador	2000	28,000	FAO, 2004
Grenada	2004	116	Eco earth
Guadeloupe	1997	2,950	FAO, 2004
Guatamala	1999	17,727	FAO, 2004
Haiti	2009	13,400	Ecoearth
Honduras	2000	78,668	FAO, 2004
Jamaica	2009	10,600	Ecoearth
Martinique	1998	1,840	FAO, 2004
Mexico	2002	882,032	FAO, 2004
Montserrat	2002	4	CCS 2002
Netherlands Antilles	2010	1,300	NetBiome
Nicaragua	1998	69,050	FAO, 2004
Panama	2000	174,435	FAO, 2004
Puerto Rico	2001	6,410	FAO, 2004
St Kitts Nevis	2002	79	CCS 2002
St Lucia	2002	179	FAO, 2004
St Vincent and the Grenadines	2002	50	FAO, 2004
Trinidad a nd Tobago	1991	7,150	FAO, 2004
Turks and Caicos Island	1991	23,600	FAO, 2004
US	2001	197,648	FAO, 2004
US virgin Islands	1999	216	FAO, 2004

3. Why Are Mangroves Important?

The importance of mangroves rests on the fact that they provide benefits in all four of the ecosystem services categories (supporting, provisioning, regulating and cultural) as defined in the Millennium Ecosystem Assessment report (2005). They contribute to coastal protection from cyclones, strong wave action, floods, sea level rise, solar UV -B radiation, greenhouse effects, and coastal erosion. They also provide the following important services for adjacent ecosystems (Ellison, 1994):

- Mangroves function as a nursery ground for many marine (crustaceans, molluscs, fin fishes, etc.) and terrestrial animals (birds, bees, and other wildlife). They provide both a habitat and a source of food for diverse communities in the forest interior and the adjacent coastal waters. There are three characteristics of mangroves that are important to their function as nursery grounds:
  - high water turbidity increases the survival rate of larvae due to reduced predation;
  - high nutrient inputs from the inflow of freshwater provide an abundant food source for zooplankton, mysids and shrimp, which, in turn are fed upon by commercially important post-larvae and juvenile species ;
  - the physical and structural complexity of the habitat provides a wide variety of niches favorable to juveniles of different species. It also provides protection from predation by more mature life stages that cannot tolerate lower salinity or cannot find prey in turbid environments.
- Mangrove root system s trap sediments creating extensive mud flats, which promote new mangrove growth. They contribute to shoreline stability and act as a sink for nutrients and contaminants. They also help to maintain offshore water quality, making it suitable for the establishment of other near-shore ecosystems such as sea-grass beds and coral reefs.
- Mangrove ecosystems provide a range of salinity conditions for a variety of organisms possessing different requirements for growth and development.
- Mangrove forests form a natural embankment and serve as a self generating barrier against storm surges and coastal flooding, thereby reducing coastal erosion.
- They supply nutrients to nearshore coastal ecosystems.
- They are of important community value providing traditional medicines, firewood, food, recreational and educational opportunities.
- Their critical economic value is associated with commercial fisheries (fish, crabs, oysters, shrimp) and tourism, as well as with prevention of coastal erosion.
- A recent survey (Donato et al., 2011) showed that mangroves are among the most carbon -rich forests in the tropics with an average biomass of 1023 Mg ha^{-1}.

4. Freshwater Flooding and Mangrove Ecosystems

Inputs of freshwater are essential to maintain the brackish water conditions typical of mangrove ecosystems. They can also influence the characteristics of mangroves by affecting salinity gradients, circulation patterns, water quality, influx of organic carbon, influx of nutrients, flushing of the ecosystem, primary productivity and the distribution and abundance of many species of plants and animals. Increases in freshwater flows can be caused by upstream activities such as deforestation, urban development, land use changes, increased surface runoff and increased precipitation. Ustach et al. (1986) reported that clearing and draining lands for agriculture resulted in a 10 % increase in freshwater flow into the Neuse River estuary in North Carolina and resulted in an increase in nutrients and turbidity. When this is coupled with extreme rainfall events that lead to increased volumes and frequency of runoff, freshwater enter mangroves as a "pulse" with the lag between rain events and runoff being greatly reduced. Excessive freshwater inflow and flooding can result in low salinity stress (to fauna and flora), increased sedimentation, density stratification, oxygen depletion, nutrient and contaminant transfer. All of these impacts can adversely affect mangroves and their associated ecotones. The relationship between freshwater flows and mangroves can be represented in a simple conceptual model as shown in Figure 5.3.

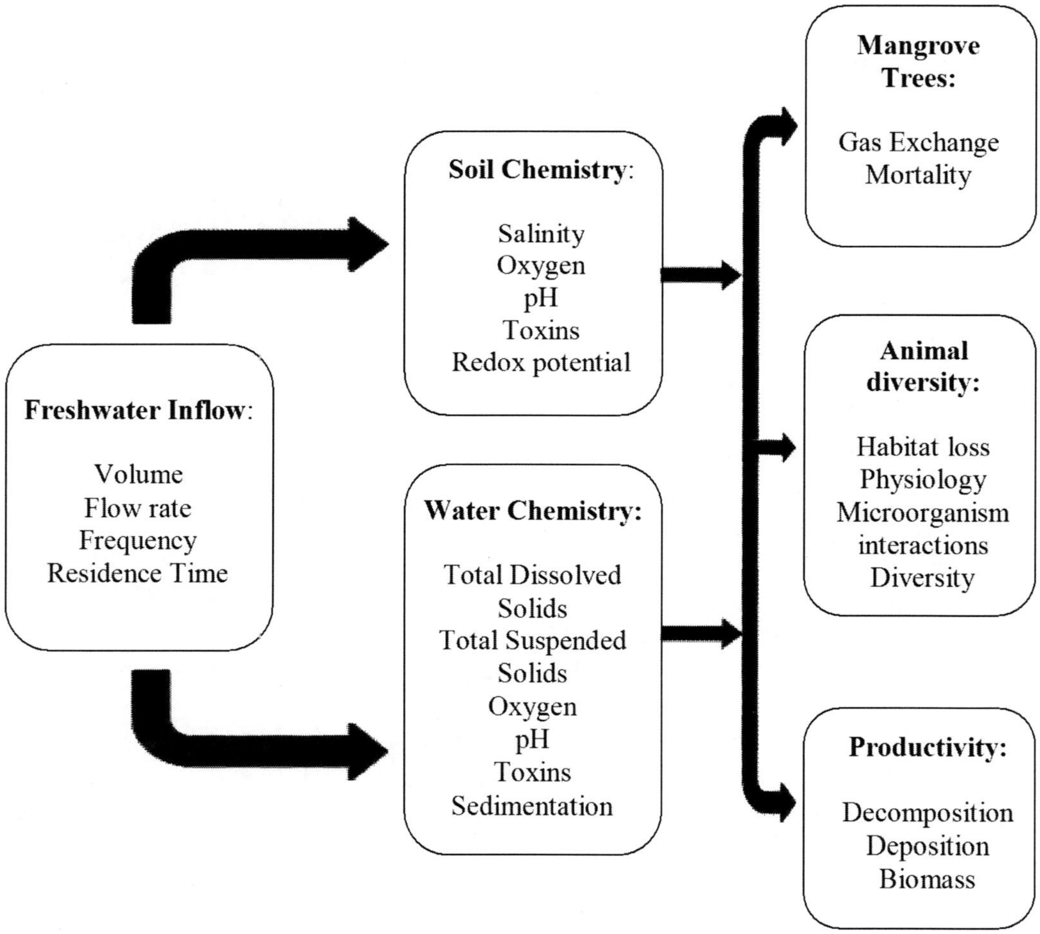

Figure 5.3. Conceptual model showing the potential linkages between freshwater flooding and mangroves.

The model gives a basic overview of the potential linkages between freshwater inflow, mangrove conditions, and biological resources. The model also indicates that the quantity, timing, dilution efficiency, flushing rates, and residence time of freshwater in mangrove systems may influence a number of important sediment and water quality characteristics. The latter can then impact upon habitats, physiology, and species composition.

4.1. Stratification

The density of seawater is greater than that of freshwater. At 20 °C, seawater has a density of about 1025 kg/m^3 while freshwater has a density of 1000 kg/m^3. Though this is only a small difference, it is responsible for driving the recirculation flow within mangroves. The main source of freshwater inflows into mangroves comes from rivers. During flood events, freshwater inflow can be relatively large in comparison to the tidal flow creating a plume of low-density (1000 kg/m^3) freshwater which flows over the higher-density seawater; resulting in a 'salt-wedge' or stratification. This stratification typically occurs in areas where the waters are deep and tidal currents and waves are not strong enough to mix the two water layers. The gradient created drives the mangrove circulation in which lighter low-salinity water at the surface moves seaward and a compensating bottom current of heavier seawater moves landward. Sediments from the freshwater settle out into the in-coming "salt-wedge" and are transported back to the upper estuary where they may be deposited (Blumberg and Oey, 1985). This brings some benefits to planktonic and juvenile animals which, rather than being swept out to sea by surface currents, are carried back into the protected and food -rich nursery. This circulation pattern plays a central role in making mangroves a suitable nursery habitat for a large number of marine organisms.

The flow of freshwater over seawater isolates the seawater from oxygen enriching processes, such as gas exchange at the air/water interface and photosynthesis by phytoplankton, leading to anoxic and hypoxic conditions in the saltwater layer. Alteration of freshwater inflows can change the circulation pattern within mangroves, thereby affecting organisms dependent upon the habitats created by the circulation patterns. When freshwater flows are high and tidal currents weak, or when there are prolonged periods with little or no wind, episodes of low oxygen (hypoxia) or absence of oxygen (anoxia) may occur. Under full flood conditions, there may be adequate mixing which results in lowered salinities and temporary flushing of saltwater out of parts of the mangrove or the entire mangrove stand. This temporarily changes the characteristics of the ecosystem to more riverine conditions. In large mangrove stands, this effect may be greatly reduced as the freshwater inputs are adequately mixed with more saline water. As low salinity water flows seawards it mixes with the underlying saltwater gradually increasing the salinity of the surface water. This is further assisted through kinetic mixing as a result of tidal flow, wind and density differences. The mixing mechanics of fresh and saltwater is usually enhanced by turbulent diffusivity and the presence of both a horizontal and a vertical density gradient (Blaise and Deleersnidjer, 2008).

4.2. Freshwater Flooding and Salinity Change

Salinity plays an important role in sustaining the biological health of mangrove ecosystems. Large inputs of freshwater can severely reduce salinity levels leading to habitat modification, a decrease in biodiversity, fish kills and a change in physiological functioning of plants and animals. All or a combination of these can result in a reduction in the health and functionality of the ecosystem. Differences in freshwater flows and associated salinity patterns have been shown to influence biodiversity and abundance of organisms (Sklar and Browder, 1998; Beever and Chamberlain, 2002; Wyda et al., 2002).

Some organisms can tolerate a wide range of salinities (i.e., are euryhaline), however, most can only tolerate a narrow range of salinities (i.e., are stenohaline). When salinity moves outside the optimal range, organisms may lose their ability to regulate internal ion concentrations and some will die. In the Caloosahatchee estuary in Florida, flooding following heavy rainfall together with freshwater releases from Lake Okeechobee were shown to depress salinity to 0-10 ppt, resulting in a reduction in salinity on oyster reefs to 0-15 ppt. This change resulted in the production of small oysters, sparse density, low spat recruitment and high mortalities (Volety & Savarese 2001; Savarese et al., 2003; Volety et al., 2003). At higher salinities (15 and 20 ppt), populations were dense, reproductive activity was high, predator numbers were low, and spat recruitment and growth increased. Salinity depression has also been observed in parts of the Caroni Swamp in Trinidad, as freshwater flooding reduced salinities to less than 10 ppt in some parts, during the wet season.

In Barbados the average salinity in the Graeme Hall mangrove lake declined from 8.4 ppt in 2002 to 1.9 ppt in 2010, a 77 % reduction which has been attributed to an inoperative sluice gate allowing large inflows of freshwater. The reduction signaled an inevitable failure of the mangrove ecosystem with a noticeable decline in bird counts and the disappearance of crab species. Freshwater flora and fauna are now dominant and the invertebrate community was found to be composed exclusively of freshwater instead of saltwater species (EEC, 2010).

Changes in salinity can impact juvenile stages of different species. Paula et al. (2003) reported complete mortality for stage 1 larva of *Parasesarma catenata* when salinity dropped to 15 ppt. Juvenile oysters (less than one year old) can survive at salinities of 5 ppt however, very little growth is observed below 5 ppt and slow growth is observed at 12 ppt (Chanley, 1958; Volety et al., 2003). Volety et al. (2003) reported that under laboratory conditions, there was high mortality (40 % to 75 %) of juvenile oysters exposed to < 5 ppt or > 35 ppt salinities, whereas very little mortality (5 %) was seen at salinities of 15–25 ppt. Low salinity also affects gametogenesis, condition index and spawning in oysters (Shumway, 1996). Salinities < 5 ppt can impair gametogenesis, whereas normal gametogenesis occurs at salinities > 7.5 ppt (Loosanoff 1953a; Loosanoff 1953b). Jyothibabu et al. (2006) similarly reported that microzooplankton abundance and biomass were highest during pre–summer monsoon and declined with the onset and peak summer monsoon in Cochin India. During the onset of the summer monsoon, the increased freshwater influx reduced the salinity of the entire study area compared with pre-summer monsoon.

4.3. Flooding and Sedimentation Processes

Mangroves flourish in sedimentary shorelines and accrete sediments at a rate of <0.5 cm per year, with a maximum of about 1 cm per year (Ellison, 1998). Moderate sedimentation is beneficial to mangroves as it acts as a source of nutrients and supplies sediment need for accretion. During flooding events, when freshwater flows are large in comparison to tidal flow, silt-laden freshwater flows into the mangrove. At the same time, sediment from the freshwater settles out into the in-coming "salt wedge" and is transported back to the upper mangrove where it can be deposited. Typically, in large intertidal estuaries, fresh water and salt water are well mixed because of the combination of shallow water and strong tidal currents.

Sedimentation in excess of the rate of accretion often occurs during flood events and may result in sediment burial, interrupting gas exchange which can lead to death of mangrove trees as a result of root smothering.. Trees may also show signs of reduced growth rate s, reduced vigor and physiological stress. The extent of impact depends on the amount and type of sedimentation, and the species involved (Ellison, 1998). Highest sedimentation rates usually occur during episodic floods which may last from a few hours to a few days, and happen only a few times each year. For example, in the Mahurangi catchment north of Auckland, New Zealand a single flood in May 1985 delivered 75 % of the estimated 20-year annual average sediment runoff to the estuary (Ellison, 1998).

Complete burial of mangrove roots (prop roots and/or pneumatophores) interrupt gas exchange, killing root tissue leading to tree death. For example, in *Avicennia spp.* mortalities have been reported when pneumatophores were buried to a depth of >10 cm (Lee et al., 1996; Ellison, 1998). The prop roots of *Rhizophora spp.,* although more tolerant than *Avicenia spp.*, can not necessarily tolerate deeper burial. For example, Grindrad (1984) reported mortality in *Rhizophora spp.* when roots were covered with 70 cm of soil and Terrados et al. (1997) reported mortality in *R. apiculata* with burial up to 32 cm. In *Rhizophora spp.*, respiration occurs through lenticels, however, the density of these decreases with increasing height above the substrate which means that under high sedimentation, gas exchange in these species can also be impaired.

Seedlings are particularly vulnerable to smothering by excessive sedimentation. Under experimental condition s, *Rhizophora apiculata* seedlings showed reduced growth and increased mortality after 8 cm of sediment burial (Terrados et al., 1997). Goodbody (1993) showed that high rates of sedimentation and eutrophication in Kingston Bay harbor, Jamaica between 1963 and 1992, led to the loss of several species of mangroves, associated ascidians and a number of other changes in the epibenthic communities. In Indonesia, Atmadja and Soerojo (1994) reported that excess sedimentation from flooded rivers caused the death of some mangrove species such as *Avicennia* spp. and *Sonneratia* spp., while Lugo and Cintron (1975) reported that mortality in mangroves following floods resulted from an interruption of root and soil gas exchange.

Ellison (1998) suggests that increased sedimentation generally may result in:

- complete burial leading to impaired gas exchange in roots and smothering of sessile organisms;
- reduced light penetration that can limit primary productivity;
- damage to fish by irritating or scouring their gills and degrading fish habitats as buried eggs becomes filled with fine particles, thus reducing available oxygen ;

- reduced success of visual predators ;
- increased toxic organic chemicals, heavy metals and nutrients that are physically and/or chemically adsorbed by sediments ; (an increase in sediment loading to the marine environment can also lead to increased deposition of these toxic substances that may result in further negative impacts such as eutrophication);
- habitat modification resulting in a loss of population and changes to community structure.

4.3.1. Freshwater Flooding and Mangrove Soil

Mangrove sediments can be described as highly anaerobic, sulphide inundated mud's that appear gray/black. The effects of salinity and particles have been discussed in previous sections. Most mangrove soils are well buffered, having a pH in the range 6-7 but some may be more acidic with a pH as low as 5. Changes in soil pH are a function of depth and seasonal changes. Marchand et al. (2004) reported that in landward margins of mangroves in French Guiana, where there were inflows of freshwater, pH values were high compared to seaward margins. Measurement of mature plurispecific and senescent mangroves at the end of a rainy season gave markedly acidic pH values, approaching 5 at the top of cores but increased to 6.7 at 40 cm depth.

Mangrove soils are typically waterlogged, giving rise to anaerobic conditions in which microbial decomposition occurs through reduction-oxidation (redox) processes, and affect the availability of the major plant nutrients nitrogen (N) and phosphorus (P). Totally anoxic soils have a redox potential (Eh) below -200 mV while oxygenated soils have a redox potential above +300 mV (Partick and Delaune 1977). At Eh less than 250 mV soil nitrates disappear and ammonium becomes the most common form of N. P is released from soil particles following iron (Fe) reduction at similar redox potentials (Reef et al., 2010 and references therein). Redox potential is very sensitive to seasonal changes, vegetal zones and depth. The lowest values are usually recorded in the upper 10 cm of cores and values decrease significantly with increasing depth. For example, Marchand et al. (2004) reported that in *R. mangle* stands, the highest Eh values were recorded at the top of cores (200 mV) and rapidly decreased to 50 mV within 5 cm. Similarly, in mixed mature forest Eh values, decreased from 200 mV at the top to 0 mV at 15 cm depth. EH values were also reportedly lowest during the wet season, ranging between 10 and 40 mV. Worley (2006) showed that *Avicennia germinans* die off, in two distinct locations, was related to unusually high rainfall in 1992 and 1995. At the die off sites, soil Eh ranged from -186 mV at 2 cm to -236 mV at 30 cm depth. This was significantly lower than at reference sites where the Eh was -88 mV at 2 cm and -155 mV at 30 cm depth. At the die off sites, surface water retention time in the wet season was 87 h while at the reference site it was only 13 h. At the die off sites there were no signs of recovery after 1 y.

Flooded mangrove soil is not only anaerobic, but can also be strongly reducing. Presence of sulphides can have negative effects on root metabolism. Sulphide concentrations in porewater and total sulphur in sediments vary with the stage of forest development, depth and species. Increased rainfall generally results in an increase in toxic hydrogen sulphide resulting from the reduction of sulphur compounds.

4.3.2. Contaminant Inputs

Floodwaters can introduce pollutants such as pesticides, metals, sewage, solid waste and nutrients into mangrove ecosystems. In December 2010, the discharge of distillery waste from rum manufacturing into tributaries upstream of the Caroni Swamp in Trinidad caused parts of the swamp to become highly anoxic resulting in high mortalities of fish and crab species. In Cuba, mangroves exposed to pesticides and other contaminants from freshwater inputs have been reported to have low rates of productivity, reduced species richness, and reduced Shannon diversity index (H') (Sanson and Rueda, 1982; Alcolado, 1984). In Barbados, the Graeme Hall mangrove wetland is reported to be disappearing due to pollution and poor water quality arising from the dumping of raw sewage into the wetland, contaminated storm water runoff, and commercial and residential pollutants from adjoining properties. A combination of these was also responsible for algal overgrowths, low water quality and fish kills (EEC, 2010).

Mangrove sediments are anaerobic and reducing, rich in sulfide, organic matter and iron and have an extraordinary capacity to accumulate pollutants (Harbinson, 1986; Silva et al., 1990; Lacerda et al., 1993; Tam and Wong, 1993, 1995a and b, 1999a, 2000). When contaminants such as heavy metals enter the mangrove environment, they are usually bound onto suspended particulates and may not at that stage pose a significant risk. Eventually, these particulates settle out and accumulate in the sediments. Although the accumulation of heavy metals in mangrove soils needs further studies, metals are known to decrease growth and respiration rates of mangroves, and are toxic to invertebrate and fish larvae. Heavy metals and pesticides may bioaccumulate in animals that inhabit mangroves (especially those closely associated with mangrove sediments), such as fish, shrimp, and mollusks. Elevated concentrations of heavy metals have been reported in mangrove sediments from various parts of the world (Harbison, 1986). Subramanian (1982) found that there was a seasonal fluctuation in the concentration of some metals (Fe, Mn, Cu, Zn) in dissolved, sediment and suspended particulate fractions of water with higher concentrations of the reactive forms during monsoon season. Oysters (*Crassostrea madresensis*) and polychaetes (*Nereis costae*) had high concentrations of metals during the monsoon season, when freshwater flooding was most frequent which further emphasizes the possibility of enrichment by freshwater input. Montgomery and Prince (1979) reported from mesocosm experiments in Puerto Rico that heavy metals leached from sewage sludge were taken up and concentrated by roots of *R. mangle*. Rojas de Astudillo et al. (2002) also reported levels of metals; Cd (0.01 ppm), Copper (5.73 ppm), Cr (0.1 ppm); Hg (0.02 ppm); Ni (0.09 ppm) and Zn (138 ppm) in *C. rhizophorae* from Trinidad.

When sediments are submerged at high tide, Fe and Mn oxides are converted into hydrated forms and provide a large surface area for reaction with metal ion s, leading to a reduction in the bio-availability of heavy metals (Tam and Wong, 1993, 1995a). Tam and Wong (1996) found that in sediments from the Futian mangroves only 2.36 % Cu, 3.79 % Zn, 0.89 % Pb and 0.09 % Cr were bio-available (i.e., could be extracted by 1 M ammonium acetate at pH 4) while Chiu and Chou (1991) reported that the exchangeable fraction of Cu, Zn, Pb, Cd, Ni and Cr in Tamshui mangrove sediments was 0.24, 1.38, 1.72, 3.17, 0.43 and 0.09 % of total heavy metals, respectively. Most mangrove tree species have developed strategies to minimize metal uptake such as metal exclusion through rhizosphere oxidation, iron plaque formation and preferential metal fixation within root tissues (Doyle and Otte,

1997; Ong Che, 1999). The oxidizing activity of *R. mangle* roots may concentrate a significant amount of the surrounding sediment metals on the root surface.

Mangroves have often been considered suitable sites for use in waste water treatment systems, as nutrient availability increases growth of mangroves (Lovelock, 2009). During flood events, excessive amounts of phosphates (derived from detergents, fertilizers and sewage) can enter mangroves from urban settlement s and agricultural runoff. Excessive amounts of phosphate can cause eutrophication and algal proliferation, which affect the natural functioning of the ecosystem. Algal growth can obstruct mangrove pneumatophores and reduce oxygen exchange, while algal mats can hinder growth of mangrove seedlings (Hogarth, 1999). High nutrient inputs can, under conditions of reduced mixing, lead to hypoxia or anoxia in bottom waters. This, in turn, may result in the death of aquatic organisms, and other undesirable consequences (Rabalais and Nixon, 2002).

Floodwaters can also introduce a range of persistent bioaccumulative toxicants (PBT) such as organochlorine pesticides and PCBs which can readily accumulate in mangrove sediments. Zonta et al. (2005) reported increased particulate matter in floodwater which was responsible for most of the heavy metal, pesticide and PBT transported into various mangrove systems in the Venice lagoon. Foster et al. (2000) also found high levels of PCBs, PAHs and OCPs in the particulate phase during flood events in Chesapeake Bay while lindane, DDT, and heptachlor have been reported in Pichavaram mangroves in India (Karthikeyan, 1988). Drouillard (2009) also reported levels of lindane (У-HCH) and PCBs in oysters from Trinidad and Jamaica. Some contaminants have also been reported in leaves, pneumatophoes, and bark from mangrove trees (Sadooni and El-Kassas 1999).

Floodwaters may also introduce pathogenic organisms into mangrove systems. These pathogens can contaminate marine organisms and pose a significant health risk if consumed. During a fish kill episode in the Windward Islands in February 2000, pathological bacteria were detected that previously had been reported only in continental freshwater systems (Caribbean Compass, 1999). This report suggested that the pathogens had been transported in sediments originating in floods in the Orinoco basin. Rampersad et al. (1999) also showed that oysters form the Caroni mangrove swamp in Trinidad were highly contaminated with *E. coli*, faecal coliforms, and *Salmonella* spp. High levels of *Vibrio* spp, which cause seafood poisoning, were also reported by Rivera et al. (1989) in shellfish from mangroves in Puerto Rico.

5. Impact of Freshwater Flooding on Mangrove Trees

The mangrove environment is characterized by two principle stress factors : salinity and hypoxia. Inundation with freshwater will usually reduce the former but can increase the latter. The direct physical impact of inundation on mangrove trees may be accompanied by locally increased salinity due to salt-wedge stratification (4.1 *Stratification*), disturbance of the substrate (*4.3 Flooding and Sedimentation processes*) or by contamination (*4.3.2 Contaminant Inputs*).

Figure 5.4. Mangrove die-back at the northern edge of the Caroni Swamp, Trinidad, West Indies.

Although there are few studies examining the impact of freshwater flooding on mangrove tree physiology, some insight can be gained from studies of saltwater flooding. Mangrove trees are adapted to withstand moderate flooding, to such an extent that flooding is required for optimal growth. Nonetheless, flooding of excessive depth, frequency or duration will reduce growth and can cause mortality. Predicting the impact of flooding is complex, with responses differing between species and between developmental stages (Krauss 2008 and references therein). Instances of prolonged, deep flooding can result in extensive tree mortalities. For both saline and freshwater flooding the principle cause is thought to be disruption of oxygen exchange between the roots and the water column or atmosphere (Jimenez and Lugo, 1985). There are a several reports of die back and mortality due to freshwater flooding. For example, Goodbody, (1993) reported significant mortality in mangrove swamps east of Port Royal in Jamaica during the rainy season. Jimenez et al. (1985) also reported massive tree mortality in Pinones-Torrecilla in Puerto Rico, an area occasionally flooded with freshwater from the Rio Grande de Loiza River. Similarly, we have observed high mortality in the northern edge of the Caroni Swamp in Beetham, Trinidad. Figure 5.4 shows the impounding of freshwater after heavy rainfall and the resultant extensive mangrove die back as well as emergence of plants favoring the reduced salinity (i.e., freshwater plants).

Although flooding is most likely to result in mortality following prolonged inundation, many physiological processes are inhibited as soon as inundation occurs. Under normal conditions, oxygen can enter root lenticels and be transported to the roots. As oxygen diffusion is > 1000 times slower in water than in air, inundation limits oxygen uptake into the soil and root lenticels (Curran et al., 1986). Soils directly adjacent to the roots may be oxidized by oxygen leaking from the root aerenchyma into the anaerobic soil. The oxidized rhizosphere around the mangrove root facilitates the conversion of ammonium (NH^{4+}) to nitrate (NO^{3-}) which can then be taken up by soil bacteria and algae. The importance of uptake of nitrate by mangrove trees is under discussion, but uptake is generally considered to be low (Reef et al., 2010).

Water depth and reduced mixing further restrict oxygen exchange in the soil and bottom water, and overtime, respiration (by plant roots and soil microorganisms) depletes the remaining oxygen. Inundation with fresh or saline water quickly depletes oxygen, e.g., experimental flooding was found to reduce soil oxygen by to up to 72 % after 20 h (Skelton and Allaway, 1996). When lenticels become inundated, oxygen concentration within the intercellular spaces of the root can drop to as little as 4–8 % (Kitaya et al., 2002). This typically causes a reduction in the oxygen partial pressure which restricts the respiratory metabolism of root cells. The low oxygen concentrations should increase oxygen influx via the aerenchyma of tissue above the floodwater, but this may not be sufficient to replenish oxygen (Curran, 1985; Curran et al., 1986; Skelton and Allaway, 1996). Under normal conditions, photosynthetically active cells at the surface of the aerial roots can feed some oxygen into the aerenchyma for root respiration; but this too, is restricted by deep flooding as these cells become shaded by the rising water (Kitaya et al., 2002).

The inundation induced hypoxia reduces root respiration, which in turn inhibits water transport, ion uptake, photosynthesis, and growth. Saltwater flooding has been shown to inhibit photosynthesis only in some cases (McKee, 1996; Naidoo et al., 1997; Chen et al., 2005; Krauss et al., 2006). Similarly, saltwater flooding generally results in reduced growth rate s, although tolerance to waterlogging differs widely between species (Pezeshki et al., 1990; Ye et al., 2003; He et al., 2007; Su et al., 2011). Allen, (2003) found that freshwater flooding enhanced growth relative to either tidal simulation or saline flooding. Naidoo et al. (1997) found that flooding with dilute river/seawater (150 mol m^3 NaCL) for 5 days had no adverse effect on photosynthesis despite the complete submergence of all pneumatophores. Pezeshki et al. (1990) found that freshwater flooding over 180 days had no effect on photosynthetic rate on a per area basis; total biomass was reduced with growth tending to be partitioned away from the leaves. In some cases, trees can avoid hypoxia by maintaining high root oxygen concentrations or by aerating the immediate rhizosphere (McKee, 1996). In the longer term, some species respond to flooding by producing more lenticels to facilitate oxygen exchange (Tomlinson, 1986; Krauss, 2008). The impact of flooding on oxygen exchange has a greater impact on the establishment of juveniles, as seedlings that lack aerial roots are unable to respire once inundated.

Under severe flooding soil conditions can develop from hypoxia into anoxia which results in the buildup of potentially toxic reduced solutes in the porewater (e.g., sulphur). Under these conditions, plants may be damaged because of an accumulation of soil phytotoxins such as reduced forms of iron or manganese, organic acids (formic and proprionic), and gases such as methane, carbon dioxide, ethylene, and hydrogen sulfide (Mckee, 1993; Youssef and Saenger, 1998). Mckee, (1993) reported that constant flooding was inhibitory to root growth in *A. germinans,* causing a decrease in both above and below ground biomass. Measurement of root gases showed that oxygen concentration decreased rapidly (by about 16 % to less than 2 %) when subjected to anaerobic conditions. The resultant oxygen deficiency causes a switch to anaerobic pathways of carbon disposal and a lowered energy status of the hypoxic roots. Some trees are able to induce changes in the soil chemistry through their above-to-below ground air transport system. This promotes oxygen loss from the roots, oxidizing rhizosphere sediments and causing oxidation of reduced constituents in the porewater. In some cases, this induces the partial precipitation of trace metals onto the root surface, creating an iron rich root coating called iron plaque (Koch and Mendelssohn 1989; Otte et al., 1989; Lacerda et al., 1992). Metal precipitation may generate concentration gradients decreasing metal bioavailability towards the roots. The co-precipitation of different metals on roots can reduce metal uptake through its retention on the iron plaque which acts as a physical barrier limiting metal uptake. Machado et al. (2005) reported that seedlings of *Avicennia schaueriana, Lagunularia racemosa,* and *Rhizophora mangle* developed an efficient exclusion of Fe, Mn and Zn through iron plaque formation.

The response of mangroves to reduced salinity is also complex. Mangroves are well adapted to life in the intertidal zone and optimal growth rate s occur in 5–75 % seawater concentrations, depending on species and developmental stage (Krauss et al., 2008). To maintain water uptake from saline substrates, mangroves restrict water loss and employ osmotic adjustment (Feller et al., 2010). Mangroves must maintain a plant water potential lower than that of their saltwater substrate, typically < -2.5 MPa (Sperry et al., 1988). Salt is actively isolated from sensitive metabolic organs by exclusion, translocation, storage, and extrusion. All of these activities carry metabolic costs, therefore inundation with freshwater would be expected to reduce salinity stress and enhance growth (Kathiresan et al., 1996). In some cases, freshwater inputs do reduce the need for osmotic adjustment and allow for increased transpiration and photosynthetic gas exchange (Naidoo 1985; Naidoo, 1993). However, any departure from the optimal salinity will reduce photosynthetic gas exchange and growth rate (Downton 1982; Burchett, 1989; Pezeshki et al., 1990; Ball et al., 1997; Ball et al., 2002; Krauss and Allen, 2003; Biber, 2006; Krauss et al., 2008). Tuffers et al. (2001) found that a site subjected to freshwater flooding had an impaired performance compared to a neighbouring site dominated by saltwater inundation. The authors attribute the reduced photosynthetic efficiency to sub-optimal salinity (salinity was < 12 ppt or approximately 35 % seawater), which was associated with lower $K+$ and N in the leaves compared to the saline site. Pezeshki et al. (1990) found that freshwater flooding reduced biomass accumulation while saline flooding increased it.

Longer-term changes in salinity can result in changes in mangrove species composition (Krauss, 2008). Where the hydrology changes such that freshwater dominates over several years, mangrove systems will gradually be outcompeted by freshwater marsh plants (Kathiresan et al., 1996; Lacerda et al., 2002) (Figure 5.4). Although there are few studies focusing on freshwater inundation, it is likely that the impact on oxygen exchange is comparable to those studies where flooding occurs without significant changes in salinity. That is, for a given mangrove system flooding outside of the optimal range inhibits growth and reproduction. Where salinity falls below a species' optimum, this will further inhibit growth and competitiveness. Mangroves are resilient to moderate flooding showing several physiological responses to the disturbance (Krauss et al., 2008), but significant mortality can occur in extreme cases particularly where flooding is combined with sedimentation (*4.3 Flooding and Sedimentation processes*).

6. Impact of Flooding on Mangrove Fauna

Freshwater inflow is one of the most significant processes responsible for shaping community structure within mangroves (Mannino and Montagna 1987; Sklar and Browder 1998). High freshwater flow can decrease the abundance of mangrove species and marine organisms that use mangroves as nursery grounds (Garcia et al., 2004). In the Caroni Swamp in Trinidad, it has been observed that root communities close to areas of freshwater input are not as abundant as those in areas away from the inflow (Figure 5.5 a & b).

Figure 5.5 a & b. Mangrove roots near sites of freshwater flows show lack of root communities.

Typically, freshwater influx increases nutrient load in mangroves, increasing chlorophyll *a* abundance and food availability for zooplankton. Freshwater can affect chlorophyll *a* in a number of different ways: (1) increasing overall chlorophyll abundance, (2) changing the location of peak chlorophyll *a* abundance and (3) decreasing chlorophyll *a* abundance and residence time. Very high flows can reduce the phytoplankton community by causing excessive flushing and insufficient residence time. In extreme cases, shifts in species dominance may occur, resulting in blooms of harmful or toxic algal species. A build up of noxious, toxic, large or otherwise inedible algae leads to inefficient zooplankton grazing and reduced trophic efficiency, thus decreasing secondary and tertiary production. Extreme fluctuation in salinity resulting from freshwater flooding often elicits stress response s or migration of some species, directly causing a reduction in biodiversity or shifts in species composition and richness. Altered freshwater delivery can influence salinity distribution within mangroves both spatially through the establishment of salinity gradients and seasonally through the relocation of the isohaline zone, which can restructure communities. Tolley et al. (2006) suggested that fish and decapod communities in southeast Florida oyster reefs were shaped partly by salinity variations, with distinct communities at different locations along a salinity gradient of individual estuaries (each experiencing different levels of freshwater inflow). Differences in community structures were greatest during wet months when salinity gradients were most expressed. Goodbody, (1961) and Beaven, (1946) showed that oyster mortality on six occasions in Chesapeake Bay coincided with heavy runoff from the Susquehanna river and consequently reduced salinity on oyster beds.

Salinity changes can also influence the physiology of mangrove fauna. Those that are not tolerant of salinity changes may die and those that are more tolerant may proliferate. Increased energetic cost and reduced performance resulting from low salinity may also occur at sites impacted by freshwater (McGraw and Reiber, 1998; Anger et al., 2000). Oxygen consumption rates increase with decreasing salinity which is confounded by the associated reduction in oxygen as floodwater enters.

Conclusion

Although freshwater flooding events are not immediately fatal to mangrove ecosystems, sustained incursion of freshwater does have the potential to destroy this valued type of habitat. Where changes in climate and land-use are seen to increase the incidents of freshwater flooding, protection of mangroves will be required.

References

Alcolado, P. (1984). Utilidad de algunos indicies ecologicos estructurales el estudion de comunidades marinas de Cuba. *Ciencias Biologicas*, *11*, 61-77.

Allen, J. A., Krauss, K W. & Hauff, R. D. (2003). Factors limiting the intertidal distribution of the mangrove species *Xylocarpus granatum Oecologia*, (2003) *135*,110–121

Anderson, C., De Silva, H., Furbert, J., Glasspool, A., Rodrigues, L., Sterrer, W. & Ward, J. (2001). Bermuda Biodiversity Country Study. Contribution 44, Bermuda.

Anger, K., Riesebeck, K. & Puschel, C. (2000). Effects of salinity on larva and early juvenile growth stages of an extremely euryhaline crab species Armases Mierii. *Hydroboligia*, *426*, 161-168.

Atmadja, W. S. & Soerojo, A. (1994). Mangrove status of Indonesia. In: Wilkinson C., S. Sudara & C. L. Ming (eds), *Proceedings Third ASEAN -Australia Symposium on Living Coastal Resources*. Australian Institute of Marine Science, Townsville, 201–207

Bacon, P. R. (1993). Mangroves in the Lesser Antilles, Jamaica and Trinidad and Tobago. *In:* Lacerda, L.D. 1993. *Conservation and sustainable utilization of mangrove forests in Latin America and Africa regions.* Vol. *2 Part I- Latin America. 155 210. Mangrove Ecosystems technical reports ITTO/ISME* Project PD114/90 (F). Okinawa, Japan. 272 pp.

Ball, M. C., Cochrane, M. J. & Rawson, H. M. (1997). Growth and water use of the mangroves *Rhizophora apiculata* and *R. stylosa* in response to salinity and humidity under ambient and elevated concentrations of atmospheric CO2. *Plant Cell Environ*, *20*, 1158–1166.

Ball, M. C. (1998). Mangrove species richness in relation to salinity and waterlogging: a case study along the Adelaide River floodplain, northern Australia. *Global Ecology and Biogeography Letters*, *7*, 73-82.

Ball, M. C. (2002). Interactive effects of salinity and irradiance on growth : implications for mangrove forest structure along salinity gradients. *Trees*, *16*, 126–39

Beaven, G. F. (1946). Effect of Susquehanna river stream flow on Chesapeake Bay salinities and history of past oyster mortalities on upper bay bars. *Contr. Chesapeake Biol. Lab.* No. *68*.

Beever, J. & Chamberlain, B. (2002). Freshwater inflow to Matlacha Pass performance measures for gator slough and Cape Coral Channel. *South Florida Water Management District Report*, 5p

Biber, P. D. (2006). Measuring the effects of salinity stress in the red mangrove, *Rhizophora mangle* L. *African Journal of Agricultural Research*, *1*, 001-004

Blaise, S. & Deleersnijder, E. (2008). A new parameterisation of salinity advection to prevent stratification from running away in a simple estuarine model. *Ocean Science Discussions*, *5*, 187-211.

Blumberg, A. F. & Oey, L. (1985). Modelling circulation and mixing in estuaries and coastal oceans. *Advances in Geophysics*, *28A*, 525-547

Burchett, M. D., Clarke, C. J., Field, C. D. & Pulkownik, A. (1989). Growth and respiration in two mangrove species at a range of salinities. *Physiologia Plantarum*, *75*, 299–303.

Carribbean Community Secretariat (CCS). (2003). *The Caricom environment in Figures 2002*, 137 p

Caribbean Compass. (1999). Fish Kill Theories Abound, but Still No Answers. *Caribbean Compass*, November 1999 http://www.caribbeancompass.com/fish.htm Accessed Jan 25th 2011

Chanley, P. E. (1958). Survival of some juvenile bivalves in waterof low salinity. *Proceedings of the National Shellfish Association*, *48*, 52-65.

Chappell, V. J. & Grindrad, J. (1984). Chenier plain formation in north Australia. In: B.G. Thom Editor, *Coastal geomorphology in Australia*. Academic Press NY, 197-231.

Chen, L. Z., Wang, W. Q. & Lin, P. (2005). Photosynthetic and physiological responses of Kandelia candel L. Druce seedlings to duration of tidal immersion in artificial seawater. *Environmental Experimental Bototany*, *54*, 256–266.

Cherrington, E. A. Hernandez, B. E. Trejos, N. A. Smith, O. A., Anderson, E. R., Flores, A. I. & Garcia, B. C. (2010). Identification of threatened and resilient mangroves in the Belize reef system. CATHALAC/WWF 33p

Chiu, C. Y. & Chou, C. H. (1991). The distribution and influence of heavy metals in mangrove forest for the tamushi estuary in Taiwan. *Soil Science and Plant Nutrition*, *37*, 659-669.

Curran, M., Allaway, W. G. & Cole, M. (1986). Root aeration and respiration in young mangrove plants (*Avicennia marina* (Fosk) Vierh) *Journal of Experimental Botany*, *37*, 1225-1233.

Curran, M. (1985). Gas movements in the roots of Avicennia marina (Forsk.) Vierh. *Australian Journal of Plant Physiology*, *12*, 97-108.

Donato, D. C., Kauffman, J. B., Murdiyarso, D., Kurnianto, S., Stidham, M. & Kanninen, M. (2011). Mangroves among the most carbon -rich forests in the tropics. *Nature Geoscience* Published online: 3 APRIL 2011 DOI : 10.1038/NGEO1123.

Downton, W. J. S., (1982). Growth and osmotic relations of the mangrove, *Avicennia marina*, as influenced by salinity. *Australian Journal of Plant Physiology*, *9*, 519–528.

Drennan, P. & Pammenter, N. W. (1982). Physiology of salt excretion in the mangrove *Avicennia marina* (Forsk.) Vierh. *New Phytologist*, *91*, 597-606.

Drew, M. C. & Dikumwin, E. (1985). Sodium exclusion from the shoots by roots of *Zea mays* (cv. LG 11) and its breakdown with oxygen deficiency. *Journal of Experimental Botany*, *36*, 55–62.

Drouillard, K. (2009). Quantitative Biomonitoring of Persistent Organic Pollutants (POPs) in Caribbean coastal zones using oysters. *Caribbean Coastal Zone Project*, 8p

Doyle, M. O. & Otte, M. L. (1997). Organism-induced accumulation of iron, zinc and arsenic in wetland soils. *Environmental Pollution*, *96*, 1-11.

ECOearth.org/article/greater_antilles_mangroves. Accessed Nov. 28th 2010

Ellison, J. C. (1994). Climate change and sea-level rise impacts on mangrove ecosystems. In: Pernetta J, Leemans R, Elder D, Humphrey S, editors. *Impacts of climate change on ecosystems and species : marine and coastal ecosystems.* IUCN, Gland, 11–30

Ellison, J. C. (1998). Impacts of Sediment Burial on Mangroves. *Marine Pollution Bulletin*, *37*, 420-422.

Environmental Engineering Consultants (EEC), Inc. (2010). Mangrove Ecosystem Assessment Graeme Hall Nature Sanctuary Barbados, 691p

FAO. (2004). Status and trends in mangrove area extent worldwide. ML Wilkie and S Fortuna, editors. *Forest resources assessment working paper*, No.63 Forest resources division FAO Rome.

FAO. (2007). The worlds mangroves 1980-2005 FAO Forestry paper No. 153. Athematic study prepared in the framework of the global forest assessment 2005.

Feller, I. C., Lovelock, C. E., Berger, U., McKee, K. L., Joye, S. B. & Ball, M. C. (2010). Biocomplexity in mangrove ecosystems. *Annual Review of Marine Science*, *2*, 395–417.

Foster, G. D., Roberts, Jr. E. C., Gruessnerb, B. & Velinskyc, D. J. (2000). Hydrochemistry and transport of organic contaminants in an urban watershed of Chesapeake bay (USA) *Applied Geochemistry*, *15*, 901-915.

Garcia, A. M., Vieira, J. P., Winemiller, K. O. & Grimm, A. M. (2004). Comparison of 1982–1983 and 1997–1998 El Ninõ Effects on the Shallow-water Fish Assemblage of the Patos Lagoon Estuary (Brazil). *Estuaries*, *27*, 905–914.

Goodbody, I. (1961). Mass mortality of a marine fauna following tropical rains. *Ecology*, *42*, 150-155.

Goodbody, I. (1993). the ascidian fauna of a Jamaican Lagoon – 30 years of change. *International Journal of Tropical Biology*, *41*, 35-38.

Harbison, P. (1986). Mangrove mud: a sink or source of trace metals. *Marine Pollution Bulletin*, *17*, 246-250.

He, B., Lai, K. T. & Harada, K. (2007). Comparison of flooding tolerance in four mangrove species in a diurnal tidal zone in the Beibu Gulf. *Estuarine*, *Coastal and Shelf Science*, *74*, 254–62.

Hogarth, P. J. (1999). The Biology of Mangroves. Oxford University Press, Oxford. 228 pp.

Jimenez, J. A., Martinez, R. & Encarnacion, L. (1985). Massive tree mortality in a Puerto Rican mangrove forest. *Caribbean Journal of Science*, *21*, 75-78.

Jimenez, J. A. & Lugo, A. E. (1985). Tree Mortality in Mangrove Forests. *Biotropica*, *17*, 177-185

Jyothibabu, R., Madhu, N. V., Jayalakshmi, K. V., Balachandran, K. K., Shiyas, C. A., Martin, G. D. & Nair, K. K. C. (2006). Impact of fresh water influx on microzooplankton mediated food web in a tropical estuary (Cochin backwaters - India) Estuarine, *Coastal and Shelf Science*, Volume *69*, Issues 3-4, 505-518

Karthikeyan, E. (1988). Organochlorine pesticides in the tropical estuary (The Vellar) mangrove (Pichavaram) and swamp (Kodikkarani) in the east coast of India. *Ph.D Thesis Annamalai University*, India 151p

Kathiresan, K., Rajendran, N. & Thangadurai, G. (1996). Growth of mangrove seedlings in intertidal area of Vellar estuary southeast coast of India. *Indian Journal of Marine Sciences*, *25*, 240-243.

Kathiresan, K. & Bingham, B. L. (2001). Biology of mangroves and mangrove ecosystems. *Advances in Marine Biology*, *40*, 81-251.

Kitaya, Y., Jintana, V., Piriyayotha, S., Jaijing, D., Yabuki, K., Izutani, S., Nishimiya, A. & Iwasaki, M. (2002). Early growth of seven mangrove species planted at different elevations in a Thai estuary. *Trees*, *16*, 150–154.

Koch, M. S. & Mendelssohn, I. A. (1989). Sulphide as a soil phytotoxine: differential responses in two marsh species. *Journal of Ecology*, *77*, 565–578

Krauss, K. W. & Allen, J. A. (2003). Factors affecting the regeneration of the mangrove, Bruguiera gymnorrhiza (L) Lamk., on a Tropical Pacific Island. *Forest Ecology and Management*, *176*, 49-60.

Krauss, K. W., Lovelock, C. E., McKee, K. L., López-Hoffman, L., Ewe, S. M. L. & Sousa, W. P. (2008). Environmental drivers in mangrove establishment and early development: A review. *Aquatic Botany*, *89*, 105–127.

Krauss, K. W., Twilley, R. R., Doyle, T. W. & Gardiner, E. S. (2006). Leaf gas exchange characteristics of three neotropical mangrove species in response to varying hydroperiod. *Tree Physiology*, *26*, 959–968

Kriedmann, P. E. & Sands, R. (1984). Salt resistance and adaptation to root -zone hypoxia in sunflower. *Australian Journal of Plant Physiology*, *11*, 287–301.

Lacerda, L. D., Conde, J. E., Kjerfve, B., Alvarez-Leon, R., Alarcon, C. & Polania, J. (2002). American mangroves. In *Mangrove ecosystems* (Lacerda, L.D. editor), 1-62 Springer-Verlag, Berlin.

Lacerda, L. D., Carvalho, C. E. V., Tanizaki, K. F., Ovalle, A. R. C. & Rezende, C. E. (1993). The biogeochemistry and trace metals distribution of mangrove rhizospheres. *Biotropica, 25*, 252-257.

Lacerda, L. D., Fernandez, M. A., Calazans, C. F. & Tanizaki, K. F. (1992). Bioavailability of heavy metals in sediments of two coastal lagoons in Rio de Janeiro, Brazil. *Hydrobiologia*, *228*, 65–70.

Liang, S., Zhou, R. C., Dong, S. S. & Shi, S. H. (2008). Adaptation to salinity in mangroves : Implication on the evolution of salt-tolerance. *Chinese Science Bulletin*, *53*, 1708–15.

Lee, S. K., Tan, W. H. & Havanond, S. (1996). Regeneration and colonization of mangrove on clay-filled reclaimed land in Singapore. *Hydrobiologia*, *319*, 23-35.

Loosanoff, V. L. (1953a). Behaviour of oysters in water of low salinity. *Proceedings of the National Shellfisheries Association* (1952), 135-151.

Loosanoff, V. L. (1953b). Lack of relationship between age of oyster or clams and the quality of their spawn *US Fish and Wildlife Service Fish Biology*, *4*, 1-2.

Lovelock, C. E., Ball, M. C., Martin, K. C. & Feller, I. C. (2009). Nutrient enrichment increases mortality of mangroves. *Plos One*, *4*(5) e5600.

Lugo, A. E. & Cintron, G. (1975). The mangrove forests of Puerto Rico and their management. In: Walsh G, Snedaker SC, Teas H (eds) *Proceedings of the international symposium on biology and management of mangroves.* University of Florida, Gainesville, 825–846

Machado, W., Gueiros, B. B., Lisboa-Filho, S. D. & Lacerda, L. D. (2005). *Trace metals in mangrove seedlings : role of iron plaque formation, 13*, 199–206.

MaGaw, I. J. & Rieber, C. L. (1998). Circulatory modification in the blue crab (Callinectes sapidus), during exposure and acclimation to low salinity. *Comparative Biochemistry and Physiology - Part A: Molecular & Integrative Physiology*, *121*, 67-76.

Mannino, A. & Montagna, P. A. (1997). Small scale spatial variations in macrobentic community structure. *Estuaries*, *20*, 159-173.

Marchand, C., Baltzer, F., Lallier-Vergès, E. & Albéric, P. (2004). Pore-water chemistry in mangrove sediments : relationship with species composition and developmental stages (French Guiana). *Marine Geology*, *208*, 361–381.

McKee, K. L. (1996). Growth and physiological responses of neotropical mangrove seedlings to root zone hypoxia. *Tree Physiology*, *16*, 883–889.

Mckee, K. L. (1993). Soil physicochemical patterns and mangrove species distribution.-reciprocal effect.? *Journal of Ecology*, *81*, 477-487.

Millennium Ecosystem Assessment (MA). (2005). Ecosystems and Human Well-Being: Synthesis. Island Press, Washington. 155pp.

Montgomery, J. R. & Prince, M. T. (1979). Release of trace metals in sewage sludge and the subsequent uptake by members of a turtle grass mangrove ecosystem. *Environmental Science and Technology*, *13*, 546-549.

Naidoo, G. (1983). Effects of flooding on leaf water potential and stomatal resistance in *Bruguiera gymnorrhiza* (L.) Lam. *New Phytologist*, *93*, 369–376.

Naidoo, G., Rogalla, H. & von Willert, D. J. (1997). Gas exchange responses of a mangrove species, *Avicennia marina*, to waterlogged and drained conditions *Hydrobiologia*, *352*, 39–47.

Naidoo, G. (1985). Effects of waterlogging and salinity on plant water relations and on the accumulation of solutes in three mangrove species. *Aquatic Botany*, *22*, 133–143.

Ong Che, R. G. (1999). Concentration of 7 heavy metals in sediments and mangrove root samples from Mai Po, Hong Kong. *Marine Pollution Bulletin*, *39*, 269-279.

Otte, M. L., Rozema, J., Koster, L., Haarsma, M. S. & Broekman, R. A. (1989). Iron plaque on roots of Aster tripolium L.: interaction with zinc uptake. *New Phytologist*, *111*, 309–317.

Paula, J., Mendes, R. N., Mwaluma, J., Raedig, C. & Emmerson, W. (2003). Combined Effects of Temperature and Salinity on Larval Development of the Mangrove Crab *Parasesarma catenata* Ortman, 1897 (Brachyura: Sesarmidae) *Western Indian Ocean Journal of Marine Science*, *2*, 57–63.

Partick, W. H. & Delaune, R. D. (1977). Chemical and biological redox systems affecting nutrient availability in the coastal wetlands. *Geoscience and Management*, *18*, 131-137.

Pezeshki, S. R., Delaune, R. D. & Patrick, W. H. (1990). Differential response of selected mangroves to soil flooding and salinity : gas exchange and biomass partitioning. *Canadian Journal of Forest Research*, *20*, 169-847.

Rabalais, N. N. & Nixon, S. W. (2002). Preface: Nutrient over-enrichment of the coastal zone. *Estuaries*, *25*, 497.

Rampersad, F. S., Laloo, S., La Borde, A., Maharaj, K., Sookhai, L., Teelucksingh, J., Reid, S., Mcdougall, L. & Adysiyum, A. A. (1999). Microbial quality of oysters sold in western Trinidad and potential health risk to consumers. *Epidemiology and Infection*, *123*, 241-250.

Reef, R., Feller, I. C. & Lovelock, C. E. (2010). Nutrition of mangroves. *Tree Physiology*, *30*, 1148-1160. Doi:10.1093/treephys/tpq048

Rivera, S., Lugo, T. & Hazen, T. C. (1989). Autecology of *vibrio vulnificus* and *vibrio parahaemolyticus* in tropical waters. *Water Research*, *23*, 923-931.

Rojas de Astudillo, L., Chang-Yen, I., Agard, J., Bekele, I. & Hubbard, R. (2002). Heavy metals in green mussel (*Perna veridis*) and Oyster from Trinidad and Venezuela *Archives of Environmental Contamination and Toxicology*, *42*, 410-415.

Rützler, K. & Feller, I. C. (1996). Caribbean mangrove swamps. Scientific American, March 1996, 94-99.

Sadooni, F. N. & El-Kassas, I. A. (1999). Mangrove as a bioindicator for environmental pollution in the coastal marine environment – review. *Qutar University Science Journal*, *19*, 137-151.

Sanson, G. G. & Rueda, R. L. (1982). Aporte de material organic del manglar alecosistema acuatico de lagunas costeras en Cuba. *Revista de Investigaciones Marinas*, *3*, 3-32.

Savarese, M., Volety, A. K. & Tolley, S. G. (2003). Influence of watershed alteration on oyster health and oyster-reef habitat : management implications for the Faka-Union and Estero Bays. *Final Report (Award # C-13252) submitted to the South Florida Water Management District*. 86 pp.

Shumway, S. E. (1996). Natural environmental factors. In: V.S. Kennedy, R.I.E., Newell and A.F. Eble, editors. *The eastern oyster Crassostrea virginica.* Maryland Sea Grant College. University of Maryland College park Maryland 467-513.

Silva, C. A. R., Lacerda, L. D. & Rezende, C. E. (1990). Heavy metal reservoirs in a red mangrove forest. *Biotropica*, *22*, 339-345.

Skelton, N. J. & Allaway, W. G. (1996). Oxygen and pressure changes measures in situ during flooding of roots of the gray mangrove *Avicenia marina* (Forsk) Vierh. *Aquatic Botany*, *54*, 165-175.

Sklar, F. H. & Browder, J. A. (1998). Coastal environmental impact s brought about by alterations to freshwater flow in the Gulf of Mexico. *Environmental Management*, *22*, 547-562

Spalding, M., Blasco, F. & Field, C. (1997). *World Mangrove Atlas.* Okinawa, Japan : The International Society for Mangrove Ecosystems: 178 pp.

Sperry, J. S., Tyree, M. T. & Donnelly, J. R. (1988). Vulnerability of xylem to embolism in a mangrove vs. an inland species of Rhizophoraceae. *Physiologia Plantarum*, *74*, 276–283.

Su, Y. M., Lin, Y. F., Jing, S. R. & Lucy Hou, P. C. (2011). Plant growth and the performance of mangrove wetland microcosms for mariculture effluent depuration. *Marine Pollution Bulletin*, *62*, 1455-1463.

Subramanian, A. N. (1982). Some aspects of cycling Iron manganese copper zinc and phosphorous in Pichavaram mangrove. Ph.D thesis Annamalai University India 252p

Tam, N. F. Y. & Wong, Y. S. (1993). Retention of nutrients and heavy metals in mangrove sediment receiving wastewater of different strengths. *Environmental Technology*, *14*, 719-729.

Tam, N. F. Y. & Wong, Y. S. (1995a). Mangrove soils as sinks for wastewater -borne pollutants. *Hydrobiologia*, *295*, 231-242.

Tam, N. F. Y. & Wong, Y. S. (1995b). Spatial and temporal variations of heavy metal contamination in sediments of a mangrove swamp in Hong Kong. *Marine Pollution Bulletin*, *31*, 254-261.

Tam, N. F. Y. & Wong, Y. S. (1996). Retention and distribution of heavy metals in mangrove soils receiving wastewater. *Environmental Pollution*, *94*, 283-291.

Tam, N. F. Y. & Wong, Y. S. (1999). Mangrove swamps as pollutant sink and wastewater treatment facility. In: Chua, T.E., Bermas, N., editors, *Challenges and opportunities in managing pollution in the East Asia n Seas.* Quezon City, Philippines, 471-483.

Tam, N. F. Y. & Wong, Y. S. (2000). Spatial variation of heavy metals in surface sediments of Hong Kong mangrove swamps. *Environmental Pollution*, *110*, 195-205.

Terrados, J., Thampanya, U., Srichai, N., Kheowvongsri, P., Geertz-Hansen, O., Boromthanarath, S., Panapitukkul, N. & Duarte, C. M. (1997). The Effect of Increased Sediment Accretion on the Survival and Growth of *Rhizophora apiculata* Seedlings. *Estuarine, Coastal and Shelf Science*, *45*, 697-701.

Tolley, S. G., Volety, A. K., Savarese, M., Walls, L. D., Linardich, C. & Everham III, E. M. (2006). Impacts of salinity and freshwater inflow on oyster -reef communities in Southwest Florida. *Aquatic Living Resources*, *19*, 371-387.

Tomlinson, P. B. (1986). The Botany of Mangroves. Cambridge University Press, Cambridge.

Tuffers, G. Naidoo, A. V. & von Willert, D. J. (2001). Low salinities adversely affect photosynthetic performance of the mangrove, *Avicennia marina. Wetlands Ecology and Management*, *9*, 225–232.

Ustach, J. F., Kirby-smith, W. W. & Barber, R. T. (1986). Effects of water shed modification on a small coastal plain estuary. In A. Wolfe, editor, *Estuarine variability*. Academic Press, Olando Florida. P 177-192.

Wyda, J. C., Deegan, L. E., Hughes, J. E. & Weaver, M. J. (2002). The response of fishes to submerged aquatic vegetation complexity in two ecoregions of the mid-atlantic bight: Buzzards Bay and Chesapeake Bay. *Estuaries*, *25*, 86-100.

Volety, A. K. & Savarese, M. (2001). Oysters as indicators of ecosystem health : determining the impacts of watershed alterations and implications for restoration. *Final Report submitted to National life Foundation, South FloridaWaterManagement District (Big Cypress Basin), and Florida Gulf Coast University Foundation.* 104 pp.

Volety, A. K., Tolley, S. G. & Winstead, J. T. (2003). Investigations into effects of seasonal and water quality parameters on oysters (*Crassostrea virginica*) and associated fish populations in the Caloosahatchee Estuary. *Interpretive Report (Award #C12412-A1) submitted to the South Florida Water management District*, West Palm Beach, Florida. 111 pp.

Worley, K. B. (2006). Relating mangrove die-off to encroaching human development in southwest Florida. NOAA Award: NA160R2407 and Florida International University, Maimi Florida 67p.

Ye, Y., Tam, N. F. Y., Wong, Y. S. & Lu, C. Y. (2003). Growth and physiological responses of two mangrove species (*Bruguiera gymnorrhiza* and *Kandelia candel*) to waterlogging. *Environmental and Experimental Botany*, *49*, 209–221.

Youssef, T. & Saenger, P. (1998). Photosynthetic gas exchange and accumulation of phytotoxins in mangrove seedlings in response to soil physico-chemical characteristics associated with waterlogging. *Tree Physiology*, *18*, 317–24

Zonta, R., Collavini, F., Zagga, L. & Zuliani, A. (2005). The effects of floods on the transport of suspended sediments and contaminants: A case study from the estuary of the Dan river (Venice lagoon Italy) *Environmental International*, *31*, 948-958. Internet resource: http://www.netbiome.org/index.php?option=com_content&view=article&id=56:netherlands-antilles-&catid=51:monographies&Itemid=30 Accessed Nov. 28th 2010 ECOearth.org/ article/greater_antilles_mangroves. Accessed Nov. 28th 2010

In: Flooding and Climate Change
Editors: D. D. Chadee, J. M. Sutherland and J. B. Agard
ISBN: 978-1-62948-317-7

Chapter 6

HEALTH EFFECTS OF FLOODING IN TRINIDAD AND TOBAGO

Dave D. Chadee*[*] *and Joan M. Sutherland
Department of Life Sciences, The University of the West Indies,
St. Augustine, Trinidad, West Indies

ABSTRACT

This chapter first provides a general introduction to the complex nature of the effects of flooding on health. It then examines international indices of poverty in Trinidad and Tobago and the national provision of the basic necessities for health and wellbeing, in particular the provision of clean piped water. Results are provided on recent research on the quality of the piped water and its vulnerability to contamination during flooding. The general level of health of the population is examined via statistics on the leading causes of death and how these may increase vulnerability to those diseases known to have the potential to be affected by flooding. Information is provided on communicable diseases including recent local research on the vector -borne diseases dengue fever, malaria and leptospirosis and discusses their potential for increase following flooding. Potential mental health effects following flooding are examined and a brief introduction is given to some of the adaptation and resilience building activities that may mitigate the effects of flooding on health.

1. INTRODUCTION

1.1. Potential Impacts of Climate Change on Health

The impacts of climate change and human activity on public health are complex. Flooding in particular can impact most sectors including agriculture, hydrology, ecosystems, communications, transport, housing and land cover which in turn can affect health. Within the Caribbean region global warming is projected to alter rainfall patterns, increase the frequency

[*] Corresponding author: Email: Dave.chadee@sta.uwi.edu; Chadee@tstt.net.tt.

and intensity of extreme events including major storm events, and increase risk of floods. Flooding can cause a range of health impacts and risks, including death and injury, contamination of drinking water, damage to infrastructure, increased populations of insect-vectors and rodents, increased mould in houses, and can create community disruption and displacement. Studies show that exposure to pathogens from sewage and unclean water can severely affect vulnerable communities with illnesses such as cryptosporidiosis, giardiasis, and norovirus (which cause diarrhoea, abdominal pain, nausea, vomiting, headache, and fever), (IPCC 2012). Flooding can have more severe effects on low- versus high-income countries (Alderman et al., 2012), however 'poverty ' or the level of development of a country is now generally recognised in terms of human development rather than income alone (UNDP, Human Development Report s) www.undp.org/en/humandev (accessed 31.12.13).

1.2. Basic Necessities for Health and Wellbeing

The health of its population is central to the wellbeing and productivity of a nation. The basic necessities for health and wellbeing are food, water, housing, sanitation (Figure 6.1); that is, the basic levels of biophysical, social and economic capital, any, or all of which may be affected by flooding. In this chapter we present an exploration of how far these basic needs are being met in Trinidad and Tobago and how flooding is likely to affect them.

1.3. Vulnerability, Adaptation and Building Resilience

The IPCC (2012) defines (i) 'vulnerability ' as "the propensity or predisposition to be adversely affected"; (ii) 'adaptation ' as: in human systems, "the process of adjustment to actual or expected climate and its effects, in order to moderate harm or exploit beneficial opportunities " and in natural systems as "the process of adjustment to actual climate and its effects; human intervention may facilitate adjustment to expected climate"; and (iii) resilience as "the ability of a system and its component parts to anticipate, absorb, accommodate, or recover from the effects of a hazardous event in a timely and efficient manner, including through ensuring the preservation, restoration, or improvement of its essential basic structures and functions" (IPCC, 2012). Adaptation measures should build the resilience of communities and individuals so that they either avoid or recover quickly from the adverse effects of flooding or other climate-change-related hazards.

1.4. The Potential Exposure to Flooding in Trinidad and Tobago

The potential exposure to flooding in Trinidad and Tobago of some of the systems on which the health of the population depends is given in Table 1.

Figure 6.1. Showing the impact of flooding on physical environment and the health and wellbeing of the affected community.

Table 6.1. Flooding-vulnerable settlements or development features located in coastal areas of Trinidad and Tobago

Settlement or development feature	% national of total
Urbanised Land	80
Industrial activities	80
Small scale industry	60
National transport arteries	50
Tourist facilities	90
Annual fish production	90

Data on flooding events collected by the meterological services of Trinidad and Tobago for the years 1985 till 2004 reveal a large increase in flooding events in years 2003 and 2004 (Figure 6.2). Data for 2010 show that there were flooding events somewhere in the country during 82 days of the year. On November 19th and 21st 2011 widespread flooding occurred in and around Port of Spain with houses being swept away by landslides and children being trapped and requiring rescue – a mother described putting her children on the table top while the water surged through her house. Another resident claimed it was the worst she had seen in her 35 years of living in the area (The Trinidad Guardian Nov 20th and 22nd, 2011).

Floods are not new to Trinidad and Tobago but occur in most rainy seasons in vulnerable parts of the country (Table 6.1). The Office of Disaster Preparedness Management has prepared flood and other hazard vulnerability maps of Trinidad and Tobago [ODPM, (2011), www.odpm.gov.tt] The questions that must be asked in a changing climate are: are floods likely to increase in frequency, extent or severity; can they be predicted more accurately; can the government and people of Trinidad and Tobago adapt to meet their potentially more damaging effects?

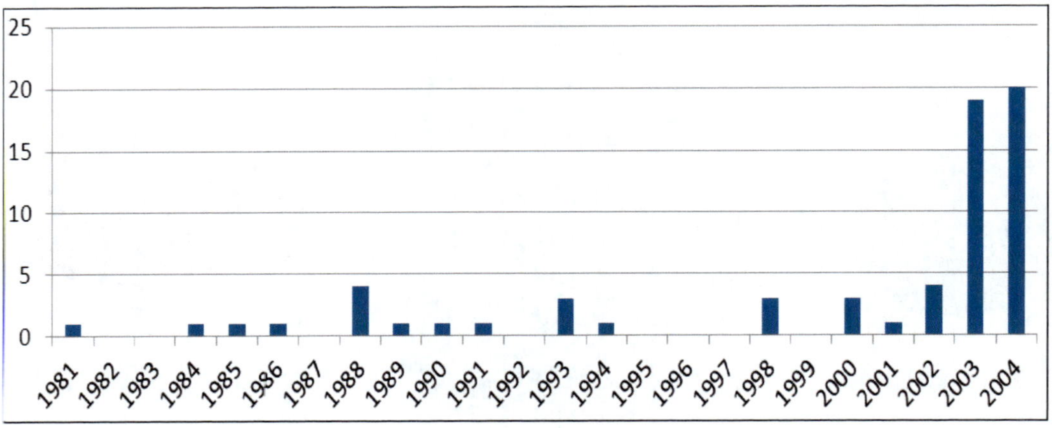

Figure 6.2. Number of Flooding Events in Trinidad and Tobago, 1981-2004.

The First National Communication (INC) on Climate Change [Environment Management Authority, (EMA) 2001] noted the difficulty of considering in isolation the effects of climate change on any one sector of national life. However an analysis carried out - informed by local consultation and expert judgement - revealed the most vulnerable sectors to be water resources and agriculture closely followed by health, coastal and marine resources, forestry and tourism.

2. Poverty, Health and Vulnerability to Natural Disasters in Trinidad and Tobago

2.1. General Poverty Indices and Vulnerability

How vulnerable a country and its people are to natural disaster s, and how easily they can adapt depend on their level of human development as well as on their geography. Recognition that human well-being rather than economic enrichment should be at the centre of development and the subsequent creation of the Human Development Report s were the achievement of Mahbub ul Haq of Pakistan and Amarta Sen of India [1]. In order to achieve international acceptance of the idea ul Haq deemed it necessary to develop a single composite Human Development Index. The current version of the Index, found in the annual development reports of the United Nations Development Program me (UNDP), consists of a composite of 3 indices: a life expectancy index, an education index and an income index. Countries around the world have now produced more than 600 Human Development Reports. In 2012 The Government of Trinidad and Tobago with the collaboration of the UNDP, produced a Human Development Atlas based on data from official statistics and national surveys covering the period from 2006 to 2010. The life expectancy index (or health index) was based on the percentage of the population suffering from at least one chronic illness and was revealed as 18.9 % of men and 25.6% of women giving a Health Index of 0.615. The Education Index was 0.742 and the Income Index 0.680 which, with the Health Index gave an

[1]http://hdr.undp.org/en/humandev (accessed 31.12.13).

overall Human Development Index of 0.677. Determination of chronic illnesses was included in the Central Statistical Office (CSO) Household Budget Survey of 2008-2009. The list of chronic illnesses included: Alzheimer's disease, arthritis, asthma, cancer, clinical mental disease, diabetes, HIV /AIDS, heart disease, hypertension, kidney disease, lupus, Parkinson's disease and sickle cell anaemia. The Human Development Atlas also contains Multidimensional Poverty Indices (MPIs) based on a Multiple Indicator Cluster Survey 2006 (MICS3) using a household survey programme developed by UNICEF that allows better data collection regarding women and children. The metrics used in the calculation of the MPIs are shown in Table 6.2. The composite Multidimensional Poverty Index for Trinidad and Tobago as well as related data is shown in Table 6.3. The overall MPI was 0.02, the percentage of the population found to be poor was 5.6% with another 0.4% vulnerable to poverty. The intensity of poverty reflects how many deprivations are experienced simultaneously. This shows that the population of Trinidad and Tobago classified as multidimensionally poor is 77,115 and a further 4,997 vulnerable to poverty.

Table 6.2. Contribution of various metrics in the calculation of the Multidimensional Poverty Index

Education		Health		Standard of Living	
Years of School	1/6	Child Mortality	1/6	Cooking fuel	1/18
School Attendance	1/6	Nutrition	1/6	Sanitation	1/18
				Water	1/18
				Electricity	1/18
				House with floor	1/18
				Asset ownership	1/18

From Central Statistical Office and UNDP (2012) Trinidad and Tobago Human Development Atlas 2012, www.cso.gov.tt, www.undp.org.tt.(accessed 31.12.13).

Table 6.3. Poverty Indices Trinidad and Tobago 2012

Multidimensional Poverty Index	0.02
Multidimensionally Poor population (%)	5.6
Population vulnerable to poverty (%)	0.4
Population experiencing severe poverty (%)	0.3
Intensity of Poverty of the Multidimensionally Poor (%)	35.1
Multidimensionally poor population	77,115
Total Population	1,241,296

From Central Statistical Office and UNDP (2012) Trinidad and Tobago Human Development Atlas 2012, www.cso.gov.tt, www.undp.org.tt

2.2. Sectorial Vulnerabilities

2.2.1. Housing and Settlement

The densest population settlement in Trinidad and Tobago occurs in the Caroni Basin. The basin is located between the northern and central ranges in Trinidad. The area is considered vulnerable to climate change and sea level rise and is home to many species of wildlife, critical life-sustaining facilities, ecosystems and the main reserves of surface and groundwater. Similarly vulnerable low-lying areas occur in the south west, east and north east of Trinidad and in the south west of Tobago (INC- EMA 2007). Frequent flooding has led residents to commence relocation efforts. Point Lisas, an industrial estate on the West coast of Trinidad, has developed into an important GDP provider. The majority of the estate is on low-lying land much of which was reclaimed. Sea level rise thus bears seriously adverse consequences for the socio-economics of Trinidad and Tobago. (CCCCC)

Large population settlements in Trinidad are located in all the foothills of the northern range – principally north and south but also east and west. Tobago is more dominated by highland and consequently a greater percentage of settlements occur on steep slopes. A combination of erodable soils and inadequately constructed houses make these settlements particularly vulnerable to storm activity. Tobago currently lies at the edge of the hurricane belt and is expected to be increasingly affected by these storms. Trinidad has been less frequently impacted by hurricanes, but projections for more intense systems could pose serious threats to both islands. (Kumarsingh and Chadee, 2008). In a UNDP Citizen Security Survey 2010, 4.5% of the population cited 'housing ' as their most serious problem.

2.2.2. Nutrition

The prevalence of undernourishment in Trinidad and Tobago in 2008 was 11% of the population [2]. The undernourished in Latin America and the Caribbean reduced from a high of 12% in 1990-92 to 8% in 2005-07 and was unchanged in 2010[3]. However, the percentage of the population who cited 'cost of food ' as their most serious problem in the Citizen Security survey 2010 was 13.9%. Any increase in flood events will impact agricultural production affecting livelihoods directly, decreasing food security, increasing price instability further discouraging production and therefore increasing poverty and hunger and hence negatively impacting health via the effects of malnutrition, loss of energy, inability to work, stress and depression and increasing vulnerability to a variety of diseases as described in chapter 7 of this book.

2.2.3. Water Supply and Sanitation

A clean water supply is regarded worldwide as essential for the maintenance of health, and wellbeing (Millennium Development Goals for 2015). During flooding events a major hazard is the overflowing of sewers into potable water systems. In Trinidad and Tobago data on the provision of potable water and on waste water treatment provided in the First Compendium on National Statistics (see Table 6.4) indicate that 90% of the population has access to potable water but only 18% has uninterrupted access and only 20% are connected to public sewers (Min of planning and environment, 2007.) By 2009 these statistics had slightly increased to 92 %, 26% and 30% respectively. The remainder of those supplied receive water on a weekly schedule, while some communities are still without a regular supply. Included among the 142 production sources are two desalination plants, the larger of which supplies 40 million gallons per day mainly to industrial clients at the Point Lisas Industrial Estate[4]. The

[2] World Bank, www.data.worldbank.org, accessed 9.1.2012

[3] FAO, www.fao.org, accessed 9.1.2012

[4] www.desalcott.com, accessed 2.12.13

second plant was commissioned in Sept 2013 and will supply 29,000 customers in the south-west of the country[5]. Of those not connected to the main 58% rely on soak-aways or septic tank s and a further 10% use pit latrines[6]...An ongoing problem in improving the reliability of the water supply is leakage from the system estimated to be around 50%[7].. Part of the difficulty in dealing with the leaks is the number of illegal buildings that have been constructed over the water mains. CDEMA (Caribbean Disaster Emergency Management Agency, www.cdema.org) includes under Environmental issues for Trinidad and Tobago: water pollution from agricultural chemicals, industrial wastes, and raw sewage.

The Beetham Waste Water Treatment Plant was commissioned in 2005, has a daily capacity of 180 million litres and treats domestic and industrial effluent from more than 350,000 inhabitants of Port of Spain and the surrounding area (www.water -technology.net). In Jan 2013 the Water and Sewerage Authority (WASA) secured a US$246 million loan from the Inter-American Development Bank (IDB) for wastewater management including the improvement of wastewater discharged into the Caroni and Cipero Rivers, integration of 8 treatment plants, construction of a new plant, and addressing the problem of the c.200 abandoned systems that have been discharging raw sewage into the environment [8]. In the recent past, several agencies and private operators unsuccessfully ran small Sewerage Treatment Plants (STP's). In 2004 most of these plants became the responsibility of WASA. (Kumarsingh and Chadee, 2008).

"Despite a high life expectancy (approaching 70 years) and a low infant mortality rate (2.4 per 1,000 births) water -related diseases are still widespread and figure among the leading causes of mortality, which is indicative of low water and sanitation service levels"[9]

Table 6.4. Data from the First Compendium of National Statistics Chapter 8 Water (Min of Planning and Development, 2007)

No of water production sources	142
Pumping stations	110
Wells	439
Surface reservoirs with a total capacity of 409 m gal	175(85in use)
Raw water reservoirs Arena Hollis Navet Hillsborough	 9.8 billion gals 1.04 billion gals 4.1 billion gals 225 million gals
Water main pipelines from 20 to 1,350 mm diam	4073.7 km
Public sewer mains	1140 km
Pop^n access to water	90%[1]
Pop^n covered by WASA waste water	20%
24/7 water supply	18%

[1]Trinidad and Tobago Human Development atlas 2012 suggests 96% of the population had access to clean water.

[5] www.news.gov.tt, accessed 2.12.13
[6] www.water -technology.net, accessed 10.1.2012
[7] www.water -technology.net
[8] Boodram, 2013, www.water -technology.net, accessed 3.12.13
[9] www.water -technology.net, accessed 3.12.13

Table 6.5. Reported Microbial and Chemical Contaminants of water Supplies

Water source/ watershed	Water supply	Chemicals & Microbial Contaminants	Reference
Urban	WASA supply to homes in San Fernando	Thermotolerant coliforms and *E. coli*	Agard et al., 2002
Rural	WASA supply to Matura, Rampenalgas, Salybia & Toco	Faecal coliforms and *E. coli*	Welsh et al., 2000
Rural	Non-WASA supply to Matura, Rampenalgas, Salibia & Toco	Faecal coliforms and *E. coli*	Welsh et al., 2000
Rural	Water drums Tamina (2005)	Soluble reactive phosphorus (SRP), phosphates, ammonia & nitrates	Hemme et al., 2009
Watershed	Hollis, Aripo & Marival	*Cryptosporidium* oocysts	Philip et al., 2007

Research conducted by UWI scientists showed that, while water samples from the Navet reservoir were negative for coliforms, 50% of samples taken at various points along the WASA supply route contained coliform bacteria indicating faecal contamination (Table 6.5 and Figure 6.3). 79% of household water samples which included those from storage vessels contained coliforms. 14.5% of the bacterial isolates from these samples produced verocytotoxins suggesting the presence of *E. coli* 0157 or other enteropathogenic *E. coli* strains. The authors (Welch et al., 2000) could only speculate as to the source of this contamination. A subsequent study (Agard et al., 2002) found increasing contamination from the reservoir (0%), to standpipe (15.2%), to household mains (53.5%) to household drinking water (80%). This increase in contamination mirrored the decline in chlorine levels. The contamination at standpipes is serious and means WASA supply is not fit for human consumption. However, most of the contamination appears to occur as a result of storage and loss of chlorine which in turn is a problem of the unreliability of the WASA supply. Hemme et al. (2009) found significant quantities of ammonium, nitrate, and soluble reactive phosphorus in water storage containers indicating contamination. Collection of rainwater supplements WASA sources in rural areas.

Cryptosporidiosis has been found in children and HIV infected adults (Tikasingh et al., 1986; Rawlins and Baboolal, 1996). Inadequate sewerage and sanitation increase the risk of water supplies being contaminated particularly in times of flooding. A study on *Cryptosporidium* contamination of three watersheds upstream of reservoirs found most contamination in the most urbanised watershed and concluded that leakage of inadequately treated sewage was responsible. The study also found significant contamination from wildlife but agricultural activity did not appear to contribute further to the contamination (Philip et al., 2007).

As well as potential contamination of drinking water, in 1996 and 1997 two bays in the N.W. peninsula of Trinidad that have both bathing beaches and marinas, were found to contain internationally unacceptable levels of faecal coliforms, particularly in the wet season (Bullock and Moonesar, 2005).

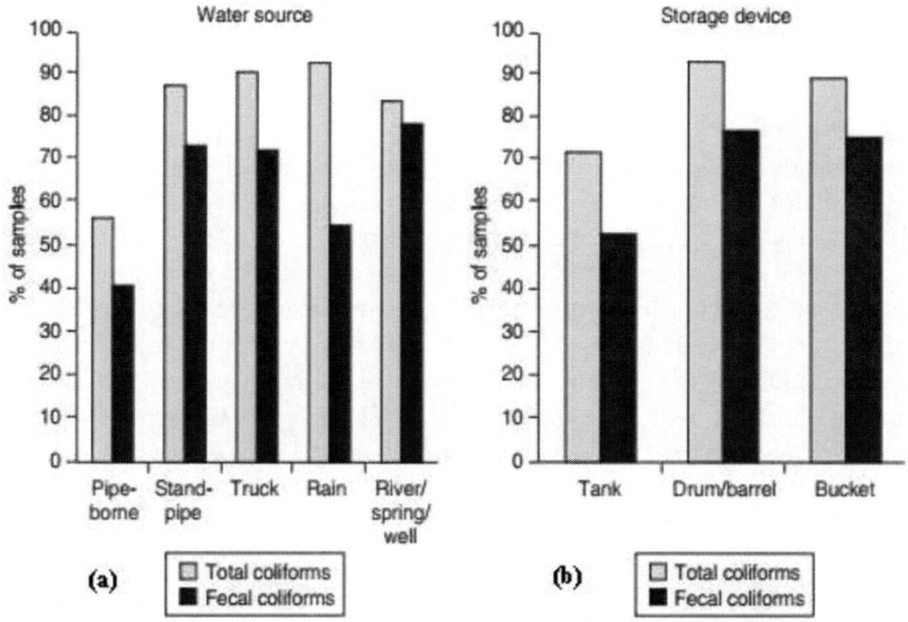

Figure 6.3. Percentage of samples contaminated with total faecal coliforms in (a) water sources and (b) storage devices, Trinidad, 1998.

2.2.4. Energy

Trinidad and Tobago Electricity Commission (T&TEC) is the sole supplier of electricity in Trinidad and Tobago. In 2008 it had 402,010 customers and supplied 7,722 GWh of electricity with a highest recorded maximum demand of 1,181 MW. The primary fuel for all generating plants in Trinidad is locally produced natural gas. While the supply is not without occasional interruptions these are now usually of short duration. The fact that fuel for generation is locally produced should mean few major disruptions in supply and relatively stable prices.

Vulnerability of major generating capacity to flooding is not known but at local level, T&TEC, the ODPM and the Red Cross of Trinidad and Tobago all publish or issue periodic hazard warnings regarding flooding and broken power lines. T&TEC have an emergency plan to restore power quickly after hurricanes.

2.2.5. Education

Trinidad and Tobago has free primary, secondary and tertiary education at government approved institutions. Adult literacy is 99%[10]. It has four universities : University of the West Indies (UWI) established in 1948 with campuses in Jamaica, Barbados and Trinidad, College of Science and Technology and Applied Arts of Trinidad and Tobago (COSTAATT) established in 2000, University of Trinidad and Tobago (UTT) established in 2004 and University of the South Caribbean (USC, formerly Caribbean Union College) a private university owned by the Seventh Day Adventist Church and established in 1927. Students gaining first class honours (equivalent to a 3.6 GPA) can apply for scholarships to study for PhDs abroad. Post disaster studies in the United States regarding adverse mental effects found

[10]www.worldbank.org, accessed 14.1.2012

that, as a group, students were particularly stressed if their studies were interrupted (Singh et al., 2014).

2.2.6. Employment

The unemployment rate in Trinidad and Tobago has been less than 10% since 2005 reaching a low of 4.5% in 2007. By 2010 it had increased to 6.4% - relatively low given the state of the world economy[11]. Average salaries are reported to be $9,657 - $22,934 for women and $12,884 - $35,368 for men[12].

A document issued to the media headlined "Government Moves to Bridge Inequality Gap" by Dookeran showed that in a 2009 survey conducted by the Central Statistical Office, measured by the international standard of inequality, T&T had a score of 0.424. This figure has increased and T&T was more unequal than in 2005, when the score was 0.39. Most developed countries tend to have a ratio between 0.24 and 0.36.

2.2.7. Security

In the Transparency International Corruption Perceptions Index report for 2011 Trinidad and Tobago received a score of 3.2 on a scale of 0 to 10 where 0 is highly corrupt and 10 very clean. Violence is also a problem and the murder rate continues to be high (2 per day for the first two weeks of 2014).

2.2.8. Healthcare

Responsibility for provision of health care services in Trinidad and Tobago was devolved from the Ministry of Health to five regional health authorities (RHAs) in 1994. The Ministry of Health continues to set policies, goals and targets for the Regions. Healthcare is delivered through 9 hospitals (1 in Tobago and 8 including a mental health hospital), 8 District Health Facilities and 95 Health Centres.

The Ministry of Health has special funds to provide external care not available locally. These are: The Children's Life Fund; Financial Assistance to Necessitous Patients or Medical Aid Programme; and the Adult Cardiac Surgery Programme.

The Ministry of Health subscribes to the WHO's Charter of Patients' Rights and Obligations. A Customer Service Implementation Unit was commissioned in 2010 to aid in the improvement of treatment to citizens. Senior citizens are given preferred treatment at all public health institutions and pharmacies[13]. The Chronic Disease Assistance Programme provides citizens with free prescription drugs and other pharmaceutical items to combat the following health condition s: diabetes, asthma, cardiac diseases, arthritis, glaucoma, mental depression, high blood pressure, benign prostatic hyperplasia (enlarged prostate), epilepsy, hypercholesterolemia, Parkinson's disease, and thyroid diseases.

There are over 250 pharmacies throughout the country that provide medications through CDAP. Despite this provision 9% of the population declared access to healthcare was their most serious problem in the UNDP Citizen Security Survey 2010 (Trinidad and Tobago Human Development Atlas, 2012).

---

[11]www.indexmundi.com accessed 10.1.12.
[12]www.payscale. com, accessed 14.1.2012
[13] From www.health.gov.tt, accessed16.1.12.

3. Leading Causes of Death

Examination of Table 6. 6, showing the leading causes of death for Trinidad and Tobago in 2003 and 2004, shows that heart disease, diabetes and cancer are the principal causes of death just as is found in developed countries. This indicates high levels of stress, poor diet and insufficient exercise. High levels of car ownership, commuting long distances to work involving hours daily in polluted highway environments leaving little time for exercise are likely to contribute to these health outcomes. Flooding incidents cause delays on highways due to flood waters on roads, landslips on roads, and frequently to erosion of roadsides often making long detours necessary, or cutting off road users entirely causing loss of income through inability to get to work or to transport goods thus adding to stress levels.

Air pollution associated with increasing traffic was identified for action in the INC. Analysis of air pollution data collected by the Environmental Management Authority (EMA) upwind of the Point Lisas Industrial Estate indicated that there is production of ground level ozone that coincides with peak traffic times (Kumarsingh and Chadee, 2008).

4. Initial Health Impacts of Flooding Events

Initial health impacts of flooding events include: drowning, crushing, fatal injuries from moving objects, and electrocution. Vulnerability to these depends on forecasting : its accuracy and whether or not warnings have been or could be acted upon

Table 6.6. Death Rates for the 10 leading causes of death in 2003 and 2004

	2003		2004	
	Rate per 100,000	% of total	Rate per 100,000	% of total
Diseases of the heart	198.1	24.9	189.5	24.8
Diabetes mellitus	111.3	14.0	106.3	13.9
Malignant neoplasm	103.2	13.0	105.2	13.8
Cerebrovascular disease	79.7	10.0	73.5	9.6
External causes	65.1	8.2	70.4	9.2
Diseases of the respiratory system	45.8	5.8	44.9	5.9
HIV/ AIDS	32.0	4.0	24.1	3.2
Diseases of the digestive system	26.0	3.3	29.7	3.9
Conditions originating in the perinatal period	22.3	2.8	14.5	1.9
Diseases of the genitourinary system	18.9	2.4	16.7	2.2

from MOH Annual Statistical Report 2004-2005.

4.1. Communicable Diseases

Those diseases that have been shown to increase following flooding events include diarrheal diseases, acute respiratory infections, malaria, leptospirosis, measles, dengue fever, viral hepatitis, typhoid fever, meningitis, as well as tetanus and cutaneous mucormycosi (IPCC AR4, 2007). Many of these diseases are shown in table 6.7 as being prevalent in Trinidad and Tobago.

4.1.1. Influenza

The most prevalent communicable disease in both 2002 and 2005 was influenza. The physical and mental stresses during and after flooding events, together with potential crowding in places of refuge are likely to hasten the spread of influenza virus.

4.1.2. Gastroenteritis

Gastroenteritis is widely reported as being the most likely negative health outcome following flooding events (Confalonieri et al., 2007). The condition of the water supply and the sewage system reported above means that flood waters will be highly contaminated with enterobacteria making the affected people highly vulnerable to gastroenteritis outbreaks. Adaptation is occurring in the form of frequent warnings and advisory notices by government and NGO agencies for emergency supplies to include bottled drinking water and bleach for cleaning. However the 11% of the population who do not have enough to eat are unlikely to be able to stockpile emergency supplies. Disaster relief agencies should distribute these in the poorest areas.

4.1.3. Acute Haemorrhagic Conjunctivitis

AHC showed an increase in prevalence rate between 2002 and 2005. This disease was predicted to be a problem in Central America and Northern South America in 1975 when Hierholzer et al. (1975) showed that, between 1969 and 1975, AHC occurred "almost exclusively in crowded coastal areas of tropical countries during hot rainy seasons. Karkai et al. (2003) and Ardkaew and Tongkumchum (2011) also found outbreaks during monsoon seasons in Nepal and Northeast Thailand respectively. Insanitary, perhaps crowded, conditions post flooding could also facilitate the spread of this disease.

4.1.4. Food-Borne Illness

Flooding events also increase the likelihood of increase in food -borne illness. The combination of increasing temperature, and vulnerability to contamination from flood waters increase the likelihood food-borne illness, particularly in urban areas where the Heat Island Effect (Xu and Chan, 2004; Balogun et al., 2010) has been shown to increase the temperature several degrees above that in adjacent open areas. Table 6.7 shows that food-borne illness in Trinidad and Tobago tripled between 2002 and 2005. No investigation into the reason for this increase has been reported adding to the uncertainty and vulnerability to further increases. Collaboration between the Health and Agricultural sectors exists in a joint Food Safety Committee. Even though responsibilities are clearly demarcated and enshrined in existing legislation, there is need to enhance institutional capacity of both sectors to carry out comprehensive food safety programmes and projects. Poor sanitation practices at food preparation establishments have contributed to periodic food illness outbreaks, especially in hotels. Preparation of local instructional materials for food handlers would assist food inspectors to more effectively communicate with the public and introduce the application of Hazard Analysis and Critical Control Points (HACCP). Inspection of licenced food provision premises is the responsibility of the Public Health Inspectorate of Environmental Health Division of the Ministry of Health. Licencing of food for sale to the public is the responsibility of the Chemistry, Food and Drug Division of the Ministry of Health.

4.1.5. Vector-borne disease

(i) *Dengue.* Table 6.7 shows high rates of dengue fever and dengue haemorrhagic fever DHF) in 2002 with lower rates of dengue and no DHF in 2005. This data is shown graphically in figures 6.4 and 6.6. Figure 6.5 confirms the seasonal character of dengue outbreaks and figures 6.5 and 6.6 show that an outbreak occurred in 2010 and 2011.

Table 6.7. Communicable disease rates (per 100,000 population estimated mid year) for 2002 and 2005

	2002	2005
Influenza	2900.4	2009
Gastroenteritis	1324.5	1685.1
Chicken pox	151.8	262.3
Acute haemorrhagic conjunctivitis	121.7	183.3
Scabies	187.6	112.7
Food borne illness	37.6	106.5
Gonococcal disease	76.3	64.5
Herpes zoster (shingles)	30.7	36.0
Dengue fever	484.7	31.1
Gonococcal disease (L)	23.9	29.8
Helminthiasis	26.5	22.2
HIV /AIDS	32.8	16.7
Tuberculosis (L)	11.4	9.7
Viral hepatitis	12.1	8.7
Leptospirosis	12.5	8.4
Meningitis	10.3	7
Salmonella	4.9	4.5
Hansen's disease	2.7	2.2
Shigella	1.2	2.2
Leptospirosis(L)	16.0	2.1
Encephalitis	0.8	1.5
Dengue fever (L)	35	0.9
Mumps	1.6	0.9
Viral hepatitis (L)	3.3	0.9
Meningococcal disease	0.3	0.1
Rubella	0.6	0.1
Tetanus	0.2	0.1
DHF	17.1	0
Malaria (L)	0.1	0
Measles	2.5	0

Epidemiological research has confirmed that outbreaks of DF are a rainy season phenomenon but also that it may be exacerbated by increasing minimum temperatures. Figure 4 shows monthly total rainfall and average minimum temperatures from 1990 and monthly dengue cases between 1997 and 2004. The increase in minimum temperature over the period is evident. The increase in dengue cases follows the increase in both rainfall and temperature by about three weeks. (Chadee et al., 2006) confirmed the seasonal nature of outbreaks with transmission taking place in the rainy season. The study also suggested that temperature may act as a trigger for the onset of DF during the rainy season. The presence of all four serotypes in the country as well as a recently imported extra vector (*Aedes albopictus*, Chadee et al., 2003) increase the vulnerability to serious outbreaks.

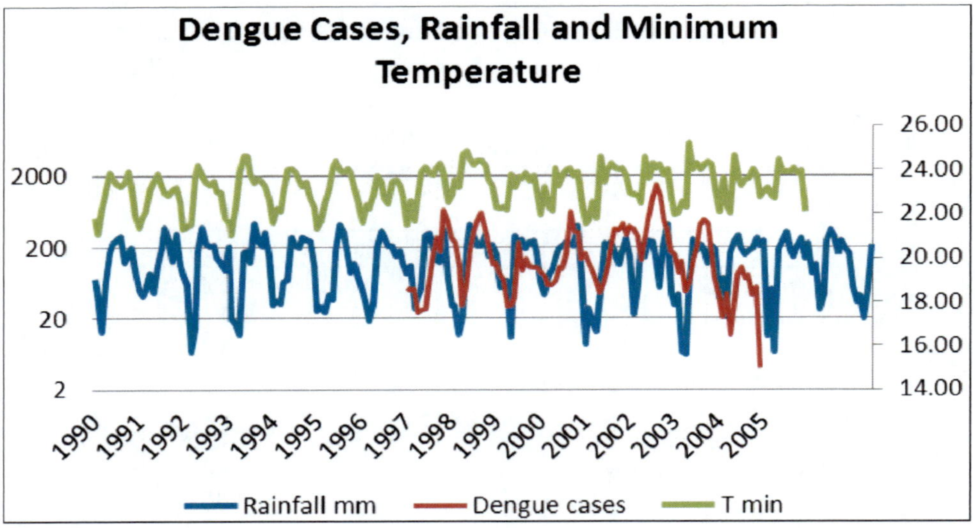

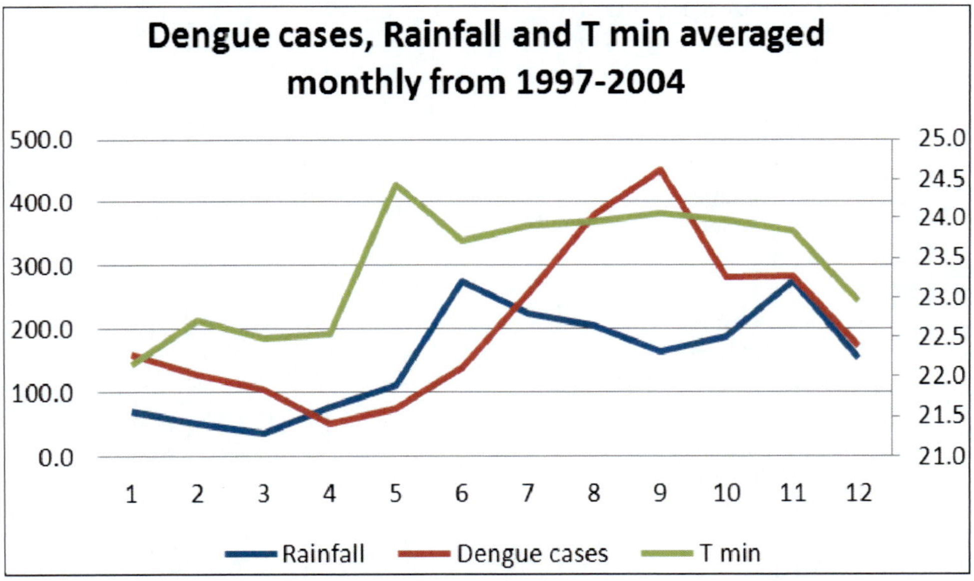

Figure 6.4. Dengue cases in relation to rainfall and temperature.

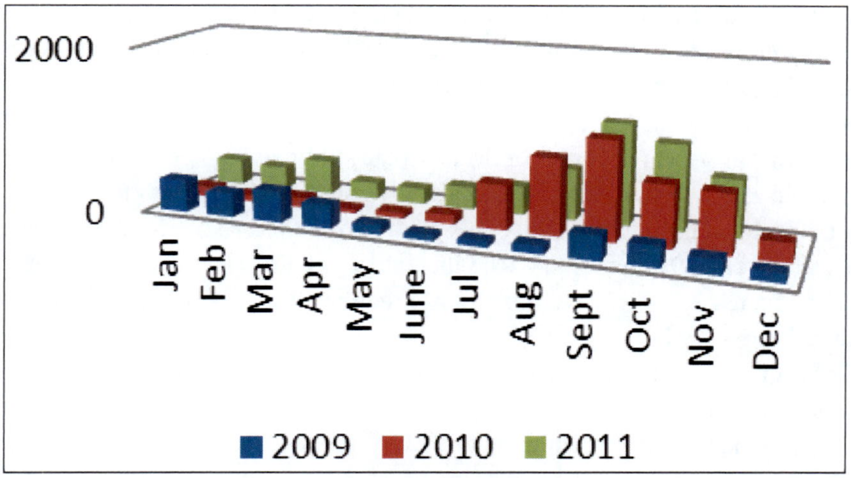

Figure 6.5. Dengue Cases in Trinidad and Tobago 2009-2011.

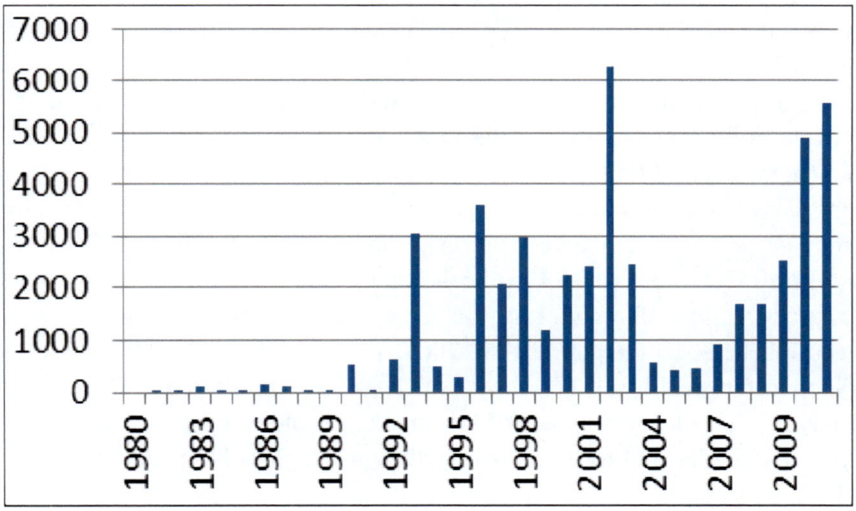

Figure 6.6. Dengue Cases in Trinidad and Tobago 1980-2011.

Ongoing research on the vector of dengue fever – the mosquito *Aedes aegypti* at the University of the West indies, St Augustine, Trinidad, The Ministry of Health Insect Vector Control Division, Trinidad and Tobago and the Caribbean Epidemiological Research Centre, Port of Spain, Trinidad, has built up a body of knowledge and experience that was recently fruitfully combined with climate and meterological experience at the University of the West Indies, Mona Campus, Jamaica resulting in the publication *Climate Change and Dengue in the Caribbean* (Chen et al., 2007). This team also contributed a chapter -*Adapting to Dengue Risk in the Caribbean* (Taylor et al., 2008) in the publication *Adaptations to Climate Change* an AIACC[1] publication edited by Leary et al., 2008).

The Insect Vector Control Division carries out an ongoing programme of surveillance, spraying and public education supported by an active research programme in collaboration with the University of the West Indies. Research into the ecology of the vectors, particularly *Aedes aegypti*, the vector of dengue fever, is of continuing international significance in its academic breakthroughs and in its practical relevance for vector control. For example identifying key premises, key containers, mosquito distribution around key premises and pre-seasonal spraying all assist in keeping cost of spraying down and in increasing efficiency and effectiveness (see Chadee, 2009).

Epidemiological research has confirmed that outbreaks of DF are a rainy season phenomenon but also that it may be exacerbated by increasing minimum temperatures. Figure 6.2 shows monthly total rainfall and average minimum temperatures from 1990 and monthly dengue cases between 1997 and 2004. The increase in minimum temperature over the period is evident. The increase in dengue cases follows the increase in both rainfall and temperature by about three weeks.

A system for detecting the seasonal increase in *Aedes* aegypti has been formulated and validated for Trinidad and Tobago but has not been included in adaptation strategies at national level. (Chadee et al., 2006, 2009). The predictable increases in *Aedes aegypti* populations and dengue cases at the beginning of the rainy season were used to develop a pre-seasonal treatment strategy that was effective in controlling both mosquito populations and dengue case numbers (Figure 6.7).

A Knowledge, Attitudes and Practice (KAP) study (Rawlins et al., 2006) that included a section on potential links between climate change and dengue found that while a majority of respondents in Trinidad and Tobago noted the wet season link with dengue fever many identified increased breeding habitats over increased temperature –related speeding up of the vector life cycle and virus replication in the mosquito host.

Heslop-Thomas et al. (2008), Taylor et al. (2008) suggested that the weakness of current dengue control strategies has been their reactive rather than adaptive nature. However an early warning system for dengue outbreaks in the Caribbean has been formulated (Amarakoon et al., 2006) but has not yet been implemented. They examined dengue cases in relation to temperature and rainfall in the Caribbean, principally Trinidad and Tobago, Barbados and Jamaica over the period 1980-2001. They found that in years with early warmer periods epidemics appeared to occur early, which was a scenario more probable in the year after an El Niño (an El Niño+1 year). A temperature index based on a moving average temperature (MAT) appeared to be most effective in gauging the potential onset of dengue (Figure 6.8).

The Insect Vector Control Division of the Ministry of Health provides services for its mosquito vector control programme based on comprehensive house to house inspections using a large work force. The relatively low level of education al attainment of some of the field staff means that intensive ongoing training in basic behaviour and communication skills for meaningful public interaction is required.

4.1.5. Vector-Borne

(ii) *Malaria* The most important vector -borne disease worldwide continues to be malaria. Trinidad and Tobago was declared malaria free in 1965. However residual cases of infection with *Plasmodium malariae* continue to be identified (Chadee et al., 1992, 1999). Of the 10 cases reported, four were found to be imported, one was a relapse case and five were described as cryptic. In the presence of other health stresses such as poor diet that may result from CC-related poor harvests, further recrudescence of malaria may occur leading to its reestablishment unless surveillance and management continue.

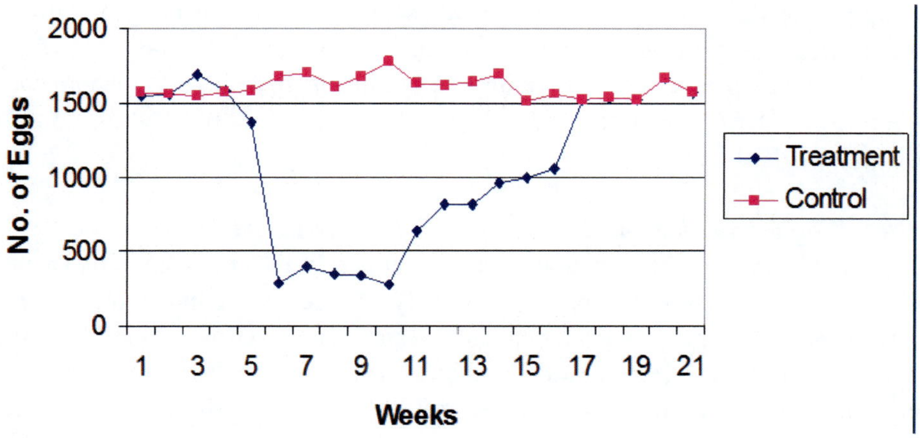

Figure 6.7. Efficacy of pre-seasonal treatment with temephos on *Aedes aegypti* ovitrap egg counts in Curepe (treatment) and St. Joseph (control), Trinidad.

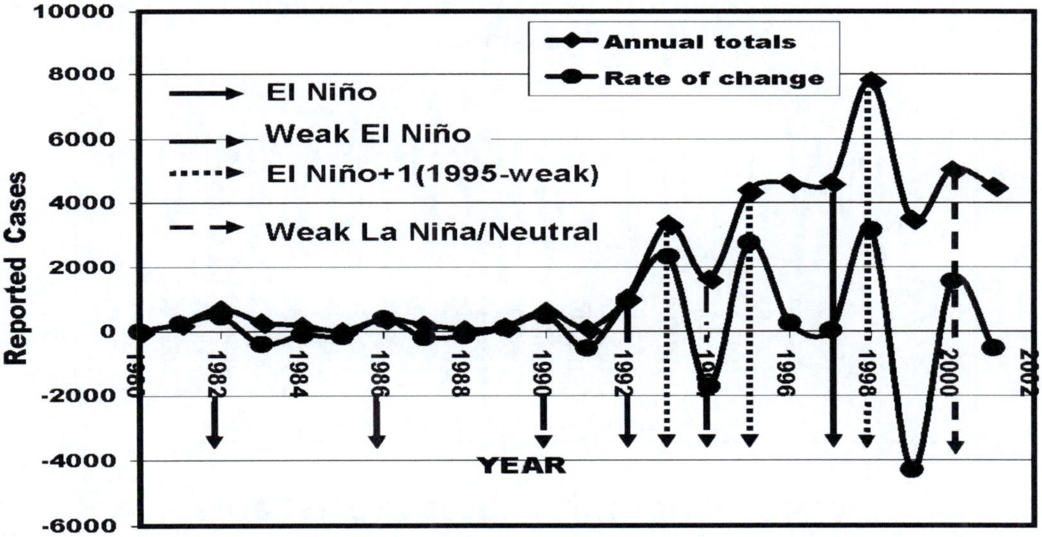

Figure 6.8.

4.1.5. Vector-borne

(iii) *Leptospirosis. Leptospirosis* is an acute febrile infection caused by bacterial species of the genus *Leptospira*. Infection is caused by exposure to water, damp soil, or vegetation contaminated with the urine of infected wild and domestic animals such as dogs and rodents (Gubler et al., 2001). Outbreaks often occur after flooding. Recent outbreaks after flooding were found in Guyana (Dechet et al., 2012), and Sao Paulo, Brazil (Miyazato et al., 2013).

Leptospirosis currently does not occur in large numbers in Trinidad and Tobago (Mohan et al., 2009). The average annual number of cases in the period 1996 to 2006 was 22 cases – 7 cases occurring in the dry season, January to June, and 15 cases in the wet season, July to December. Monthly rainfall distribution and leptospirosis cases over the period 1996 to 2006 and the monthly average number of cases and rainfall over the period are shown in figure 6.9. There is only a slight lag in case occurrence/reporting after the onset of the rainy season suggesting that infections occur immediately after the rains begin. GIS of case locations showed that most occurred in sub-urban areas on residential land. A KAP survey revealed a low level of awareness of the disease but dissatisfaction with government rodent control services. A SWOT analysis revealed lack of rodent control resulted from insufficient funding (Mohan, 2010).

4.1.6. Other Communicable and Emerging and Re-Emerging Diseases

Cases of Meningococcal infections continue to be registered with no evident trend suggesting epidemiological outbreaks. The majority of cases are found in the under 19 age group. Acute haemorrhagic conjunctivitis showed an increase during 2003. Salmonellosis, shigellosis, influenza and viral hepatitis B showed increasing registry from 2002 to 2004. Prevalence rates for Hansen disease have remained unchanged since 1999. No cases of lymphatic filariasis (LF) have been reported since 1982. A national survey conducted in 2002 in school children 6-12 using ICT (immunochromatographic card tests) provided evidence of the interruption of LF transmission in Trinidad (Rawlins et al., 2004).

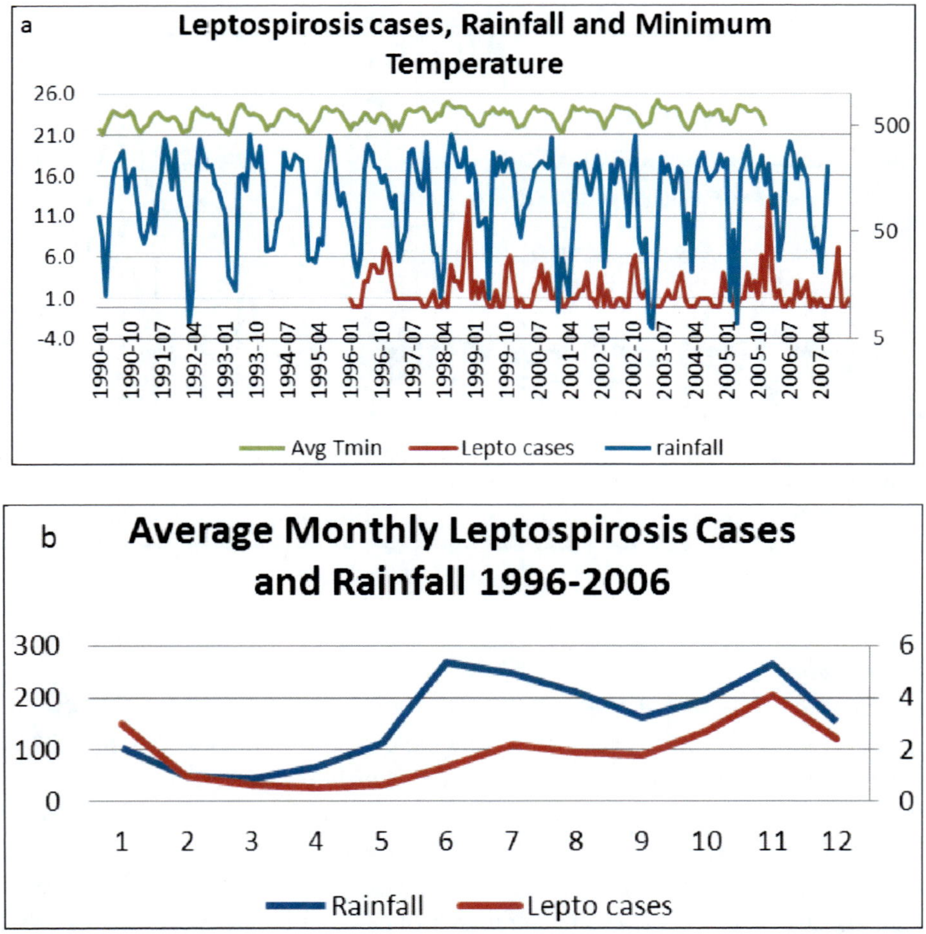

Figure 6.9. Leptospirosis cases (a) in relation to rainfall and minimum temperature from 1996 to 2007 and (b) in relation to monthly rainfall.

4.2. Mental Health

The severity of the effects of flooding events on the mental health of an individual depends on a large number of variables: the severity of the event experience on the person, their family, their property; on the person's age, sex, social status, previous experience of floods and preparedness e.g., whether insured; the duration of the flood and whether evacuation has been necessary. A large research project funded by the European Community (FlOODsite, 2009) produced data on the impacts of flooding and the factors influencing human behaviour before, during and after flood events. In defining "social capital " the researchers accept the view of Bourdieu (1986) who defines it as the "aggregate of the actual or potential resources which are linked to possession of a durable network of more or less institutionalised relationships of mutual acquaintance and recognition ". The data showed that reactions were population - and location-specific and also complex and frequently counter intuitive e.g., the elderly were not necessarily more severely affected than other age groups. A recent post-flood study of older adults showed that one in six reported post-traumatic stress disorder (PTSD) symptoms that might require clinical attention (Bei et al., 2013). Students who had their work disrupted by a major flood in Midwestern USA were four times more likely to report PTSD symptoms than those that had not had their work disrupted (Peek-Asa et al., 2012). Another group who may require post-disaster psychological support are the government or other agency first responders (Kitamura et al., 2013). In two flood prone areas of Trinidad, a KAP study was conducted which suggested that further studies, on mental health issues be conducted (Singh and Chadee, Chapter 7).

4.3. Summary of Health Effects of Flooding

A summary of the main effects of flooding on health in Trinidad and Tobago is given in table 6.8. These are divided into direct and indirect effect s and also short and longer term effects.

Table 6.8. Effects of flooding on health in Trinidad and Tobago

	Short Term	Longer Term
Direct	Direct injury or death -from flooding, landslides or hurricanes (particularly Tobago) Increase in hunger and malnutrition from crop failure and increased food prices Increase in respiratory morbidity and mortality - from increased dust and ozone	Psychological effects – anxiety and depression - from increasing insecurities, post- traumatic stress Increased heart disease from decreasing quality of life and resulting increased stress Trauma from permanent loss of homes from inundation through sea level rise or increased seasonal flooding or landslides
Indirect	Increase in water- borne diseases, diarrhoeal diseases, vector- borne diseases: dengue (high risk) ; leptospirosis (high risk); lyme disease(?); malaria (medium risk).	Long term physical and psychological conditions resulting from disease and increasing insecurity of water, food and shelter.

5. Adaptation and Resilience Building

5.1. The Adaptation Partnership

The Adaptation Partnership (www.adaptationpartnership.org) was formed in 2010 in "response to a recognised need for development practitioners to share information and lessons on adaptation efforts" and "to catalyse action and foster communication among the various institutions and actors engaged in the effort to scale up adaptation and resilience around the world, particularly in the context of fast start finance". To this end they have produced a review of current and planned adaptation action in twelve countries of the Caribbean including Trinidad and Tobago (Madeiros et al., 2011).

5.2. The Caribbean Disaster Emergency Management Agency

It is 20 years since CARICOM set up the first regional disaster management agency and in Dec 2012 (5th-9th) the 6th Caribbean Conference on Comprehensive Disaster Management took place in Trinidad. In her remarks at the Closing Ceremony, Andria Grosvenor, Technical Manager, Preparedness and Country Support, CDEMA (Caribbean Disaster Emergency Management Agency) and 2011 Conference Chair, indicated that "although progress has been made over the past twenty years, a lot more needs to be done in the field of comprehensive disaster management (CDM) in the region. She opined that "political awareness has grown but this must be transferred into action." CDM is the regional contribution to the United Nations ' International Strategy for Disaster Reduction's (UNISDR) Hyogo Framework for Action 2005-2015, a blueprint for global action in Disaster Risk Reduction[14].

5.3. Health Sector Preparation in Trinidad and Tobago

The ministry of health and Regional Health Authorities are in the process of mainstreaming health promotion initiatives and integrated management of non-communicable diseases NCDs including health promotion in school s. Policies are in place for the majority of these issues but require further reinforcing and compliance at all levels together with improved monitoring and surveillance systems.

Environmental health services within the Ministry of Health remain focused on inspections and registrations for food safety and the monitoring for mosquito vector control and general sanitation within communities and institutions. Environmental Health is an integral part of the Ministry of Health, although eventual decentralization to the Regional Health Authorities (RHAs) under the Health Sector Reform Programme was planned but under the current government this may be reviewed.

Public perceptions found via KAP studies in relation to CC (Rawlins et al., 2006) ranked sector importance in the following order: health, water resources, agriculture, biodiversity and coastal degradation. Health issues were ranked: food -borne diseases, water-borne diseases,

[14] www.cdema.org, accessed 13.1.2012

heat stresses, respiratory diseases and vector -borne diseases, perhaps confirming the existing effects on health of poor air and water quality referred to above. In spite of their knowledge of dengue and its causes water-related illnesses are considered more serious suggesting that, until people have a reliable potable water supply they will continue to treat the threat of vector-borne diseases as less important and attempts to involve communities in environmental clean-up programmes will continue to fail.

The processes of adaptation and resilience building are iterative processes between the major actors involved featuring alternately and/or simultaneously national and or local level activities; national disaster organisations (NDOs) non-government organisations (NGOs), community and district groups and councils, and researchers each taking leading or secondary roles depending on the stage of the process (Figure 6.10). Trinidad and Tobago may be ahead of other Caribbean countries in disaster preparedness development in that the Office of Disaster Preparedness and Management, a division of the Ministry of Security, is already the principal first responder following disasters – and has been involved in all the recent flood relief activities in the country. In the severe flooding which occurred in the Port of Spain area several notable circumstances occurred: the mayor of Port of Spain, Mr. Lee Sing, noted that "shortly after the rain had stopped communities had come out to clean the streets and assist neighbours in affected areas" indicating considerable social capital in the area; ODPM, workers from the three affected regional councils, groups from CEPEP[15] and the Ministry of Works and Infrastructure were involved in clean- up operations ; the ODPM issued an advisory on the use of bleach in cleaning up homes and possessions; and also indicated that the Ministry of the People had been mandated to work with affected residents to complete damage and loss assessments (Guardian, 22.11.11). These prompt and coordinated activities help to build confidence and reduce the likelihood of the flood victims suffering psychological after effects.

Table 6.9. A proposed adaptive strategy for the reduction of climate change effects on health

Strategy	Action	Impact
Preseasonal Interventions	Cleaning of water canals	Flooding management
	Treatment of water	Dengue and vector control
	Source reduction programme	Environmental sanitation and mosquito control
	Refresher training	Capacity building
	Planning – monitoring and evaluation	Assessment of impact
	Community participation	Community Empowerment

One of the most important actors in helping communities to increase their resilience to disaster is the Trinidad and Tobago Red Cross Society; with the help of the European Commission Humanitarian Aid Department (ECHO) and the Finnish Red Cross Society they have conducted workshops in 14 communities in 7 disaster-prone regions. The training has included Community Based Health and First Aid (CBHFA) and Vulnerability and Capacity

[15] Community –based Environmental Protection and Enhancement Programme

Assessment (VCA) in which communities assess their vulnerabilities and learn how to mitigate and plan and respond to disasters. They also publish regularly in the national newspapers instructions on how to prepare for disasters.

Figure 6.10. Summary of the tripartite interactions required for adaptation and resilience building.

Conclusion

Evidence is provided which shows an association between climate change, higher temperatures, and flooding frequency with respect to human health. Increases in outbreaks of water -, food - and vector -borne diseases are *very likely* unless all the relevant stakeholders such as communities, national disaster organizations, non-governmental organizations, policy makers, researchers and educationalists increase their adaptation efforts.

References

Agard, L., Alexander, C., Green, S., Jackson, M., Patel, S. & Adesiyun, A. (2002). Microbial quality of water supply to an urban community in Trinidad. *Journal of Food Production*, *65*, 1297-1303.

Amarakoon, D., Chen, A., Rawlins, S. C., Taylor, M. & Chadee, D. D. (2006). Retrospective study. In: A.A. Chen, DD. Chadee and SC.Rawlins (eds.) Climate Change Impact on Dengue: The Caribbean Experience. Climate Studies Group, Mona, University of the West Indies, Mona Kingston, Jamaica in assocation with International START Secretariat 2000 Florida Ave NW. Washington DC 20009 USA

Ardkaew, J. & Tongkumchum, P. (2011). Modeling the hospital burden of common infectious diseases with application to northeastern Thailand. *Asia Pacific Journal of Public Health*, *23*, 703-711.

Balogun, A. A., Balogun, I. A. & Adeyewa, Z. D. (2010). Comparisons of urban and rural heat stress conditions in a hot-humid tropical city. *Global Health Action*, *3*, 10.3402/gha.v3i0.5614.

Bei, B., Bryant, C., Gilson, K. M., Koh, J., Gibson, P., Komiti, A., Jackson, H. & Judd, F. (2013). A prospective study of the impact of floods on the mental and physical health of older adults. *Aging and Mental Health*, *17*, 992-1002.

Bourdieu, P. (1986).The forms of capital. In: Richardson, JG (Ed), Handbook of Theory and Research for the Sociology of Education. Greenwood, New York, (241-258).

Bullock, C. A. & Moonesar, I. (2005). Potential sourcxes of bacteriological pollution for two bays with marinas in Trinidad. *Rev. Biol.Trop.*, *53*, Suppl 1, 91-103.

Chadee, D. D. (2009). Impact of pre-seasonal focal treatment on population densities of the mosquito *Aedes aegypti* in Trinidad, West Indies : A preliminary study. *Acta tropica*, *109*, 236-340.

Chadee, D. D., Beier, J. C. & Doon, R. (1999). Re-emergence of *Plasmodium malariae* in Trinidad, West Indies. *Annals of Tropical Medicine and Parasitology*, *93*, 467- 475.

Chadee, D. D., Hong Fat, F. & Persad, R. C. (2003). First record of *Aedes albopictus* in Trinidad, West Indies. *Journal of the American Mosquito Control Association*, *19*, 438-439.

Chadee, D. D., Le Maitre, A. & Tilluckdharry, C. C. (1992). Reappearance of vivax malaria in Trinidad, West Indies. *Transactions of the Royal Society of Tropical Medicine & Hygiene*, *86*, 485.

Chadee, D. D., Shivnauth, B., Rawlins, S. C. & Chen, A. A. (2006). *Dengue Fever and Climate Variability: A Prospective Study in Trinidad and Tobago.* In: AA Chen; DD Chadee & SC Rawlins (eds.) Climate Change Impact on Dengue: The Caribbean Experience. Climate Studies Group, Mona, University of the West Indies, Mona Kingston, Jamaica in assocation with International START Secretariat 2000 Florida Ave NW. Washington DC 20009 USA

Chadee, D. D. & Wilkerson, R. C. (2005). *Anopheles triannulatus* (Neiva and Pinto): a new *Anopheles* record from Trinidad, West Indies. *Journal of the American Mosquito Control Association*, *21*, 316-317.

Chadee, D. D. & Wilkerson, R. C. (2006). Ecology of the malaria vector, *Anopheles (Nyssrhynchus) marajoara* Galvo and Demasceno in Trinidad, West Indies. *Journal of the American Mosquito Control Association*, *22*, 22-28.

Confalonieri, U., Menne, B., Akhtar, R., Evi, K. L., Hauengue, M., Kovats, R. S., Revich, B., & Woodward, A. (2007). Human Health. *Climate change 2007: Impacts, Adaptation and Vulnerability. Contribution of Working Group II to the Fourth Assessment Report of the Intergovernmental Panel on Climate Change*, ML Parry; OF Canziani; JP Palutikof; PJ van der Linden & CE Hanson, Eds., Cambridge University Press, Cambridge, UK, 391-431.

Dechet, A. M., Parsons, M., Rambaran, M., Mohamed-Rambaran, P., Florendo-Cumbermack, A., Persaud, S., Baboolal, S., Ari, M. D., Shadomy, S. V., Zaki, S. R., Paddock, C. D., Clark, T. A., Harris, L., Lyon, D. & Mintz, E. D. (2012). Leptospirosis outbreak following severe flooding : a rapid assessment and mass prophylaxis campaign; Guyana, January-February 2005. *Plos One*, *7* (7), e39672.doi: 10.1371/journal.pone.0039672. Epub2012 Jul 9.

Environment Management Authority, Government of Trinidad and Tobago. (2001). First National Communication of the Republic of Trinidad and Tobago under the United Nations framework Convention on Climate Change.

FLOODsite. (2009). Flood risk assessment and flood risk management. An introduction and guidance based on experiences and findings of FLOODsite (an EU -funded Integrated Project) Deltares/Delft Hydraulics, Deft, the Netherlands. www.floodsite.net

Gowrie, M. N. (2008). Association between pediatric asthma admissions, Sahara dust, meterological parameters, and airborne pollen content in Trinidad and Tobago. MPhil Thesis, University of the West Indies, St. Augustine, Trinidad.

Gubler, D. J., Reiter, P., Ebi, K. L., Yap, W., Nasci, R. & Patz, J. A. (2001). Climate variability and change in the United States : potential impacts on vector - and rodent-borne diseases. *Environmental Health Perspectives*, *109* Suppl 2, 223-233.

Hemme, R. R., Tank, J. L., Chadee, D.D. & Severson, D. W. (2009). Environmental conditions in water storage drums and influences on *Aedes aegypti* in Trinidad, West Indies. *Acta Tropica*, *112*, 59-66.

Heslop-Thomas, C., Bailey, W., Amarakoon, D., Chen, A. A., Rawlins, S., Chadee, D. D., Crosborne, R., Owino, A., Polson, K., Rhoden, C., Stennett, R. & Taylor, M. (2008). In: *Climate Change and Vulnerability*. Edited by: N Leary; C Conde; J Kulkarni; A Nong; & J Pulbin, Eds (2008). Earthscan, London. Pp 428.

Hierholzer, J. C., Hilliard, K. A. & Esposito, J. J. (1975). Serosurvey for "acute hemorrhagic conjunctivitis " virus (Enterovirus 70) antibodies in the Southeastern United States, with review of the literature and some epidemiologic implications. *American Journal of Epidemiology*, *102*, 533-544.

IPCC. (2012). Summary for Policymakers. In: Managing the Risks of Extreme Events and Disasters to Advance Climate Change Adaptation [Field, C.B., V. Barros, T.F. Stocker, D. Qin, D.J. Dokken, K.L. Ebi, M.D. Mastrandrea, K.J. Mach, G.-K. Plattner, S.K. Allen, M. Tignor, and P.M. Midgley (eds.)]. A Special Report of Working Groups I and II of the Intergovernmental Panel on Climate Change. Cambridge University Press, Cambridge, UK, and New York, NY, USA, 1-19.

Karki, D. B., Shrestha, C. D. & Shrestha, S. (2003). Acute haemorrhagic conjunctivitis : an epidemicin August/September 2003. *Kathmandu University Medical Journal (KUMJ)*, *1*, 234-236.

Kitamura, H., Shindo, M., Tachibana, A., Honma, H. & Someya, T. (2013). Personality and resilience associated with perceived fatigue of local government employees responding to disasters. *Journal of Occupational Health*, *55*, 1-5.

Kumarsingh, K. & Chadee, D. D. (2008). Country Profile on Climate Change and Health Trinidad and Tobago. Report to the Pan American Health Organization/World Health Organisation 27th March 2008.

Madeiros, D., Hove, H., Keller, M., Echeverria, D. & Parry, J-E. (2011). Review of current and planned adaptation action: The Caribbean. www.adaptationpartnership.org

Miyazato, K. E., Fonseca, A. L., Caputto, L. Z., Rocha, K. C., Azzalis, L. A., Junqueira V. B., Pereira, E. C., Chaves, L. C., Feder, D., Corazzini, R., De Abreu, L. C., Valenti, V. E., Batista Lacerda, S. N., Goulart, F. C. & Fonseca, F. L. (2013). Incidence of Leptospirosis infection in the East Zone of Sao Paulo City Brazil. *International Archives of Medicine*, *6*, 23.

Mohan, A. R. M. (2010). Assessment of the current status of human leptospirosis and the state of rodent control in Trinidad and Tobago using an evidence -based approach. PhD thesis University of the West Indies, St. Augustine.

Mohan, A. R. M., Cumberbatch, A., Adesiyun, A. A. & Chadee, D. D. (2009). Epidemiology of human lepospirosis in Trinidad and Tobago, 1996-2007: A retrospective study. *Acta Tropica*, *112*, 260-265.

Patt, A. G., Tadross, M., Nussbaumer, P., Asante, K., Metzger, M., Rafael, J., Goujon, A. & Brundrit, G. (2010). Estimating least-developed countries ' vulnerability to climated-related extreme events over the next 50 years. *Proceedings of the National Academy of Sciences*, *107*, 2333-1337.

Peek-Asa, C., Ramirez, M., Young, T. & Cao, Y. (2012). Flood-related work disruption and poor health outcomes among university students. *Prehospital Disaster Medicine*, *27*, 503-508

Philip, D. A. T., Rawlins, S. C., Babooll, S., Gosein, R., Goddard, C., Legall, G. & Chinchamee, A. (2007). Relative importance of the various environmental sources of *Cryptosporidium* oocysts in three watersheds. *Journal of Water and Health*, *6*, 23-34.

Ramkisson, S. (2010). Evaluation of sampling methods for enumeration of microorganisms associated with Sahara dust in Trinidad and Tobago. MPhil Thesis, University of the West Indies, St. Augustine, Trinidad.

Rawlins, S. C., Siung-Chang, A., Baboolal, S. & Chadee, D. D. (2004). Evidence for the interruption of transmission of lymphatic filariasis among school children in Trinidad and Tobago. *Transactions of the Royal Society of Tropical Medicine and Hygiene*, *98*, 473-477.

Rawlins, S. C., Bailey, W., Chadee, D. D. & Heslop-Thomas, C. (2006). Knowledge, Attitude and Practices (KAP) studies in Trinidad and Tobago, St Kitts-Nevis and Jamaica. In: AA. Chen; DD. Chadee; & SC Rawlins (eds.) Climate Change Impact on Dengue: The Caribbean Experience. Climate Studies Group, Mona, University of the West Indies, Mona Kingston, Jamaica in assocation with International START Secretariat 2000 Florida Ave NW. Washington DC 20009 USA.

Rawlins, S. C. & Baboolal, S. (1996). *Cryptosporidium* in some Caribbean children and an evaluation of the efficacy of tests for its laboratory diagnosis. *West Indian Medical J.*, *45* (Suppl 2), 17.

Tamerius, J., Nelson, M. I., Zhou, S. Z., Viboud, C., Miller, M. A. & Alonso, W. J. (2011). Global influenza seasonaliy: reconciling patterns across temperate and tropical regions. *Environmental Health Perspectives*, *119*, 439-445.

Taylor M. A., Chen, A., Rawlins, S., Heslop-Thomas, C., Amarakoon, D., Bailey, W., Chadee, D. D., Huntley, S., Rhoden, C. & Stennett, R. (2008). In *Climate Change and Adaptation* Edited by: N. Leary, J. Adejuwon, V. Barros, I. Burton, J. Kulkarni & R. Lasco. Earthscan, London. Pp 381.

Tikasingh, E. S., Smith, S., Poon-King, T. & Bratt, D. (1986). *Cryptosporidium* in man in Trinidad and Tobago. *Trans. R. Soc. Trop. Med. Hyg.*,*80*, 998-999.

Water-technology.net www.water -technology.net/projects/construction Water Plant Construction and Rehabilitation, Water and Sewerage Authority (WASA), Trinidad and Tobago

Welch, P., David, J., Clarke, W., Trinidade, A., Penner, D., Bernstein, S., McDougall, L. & Adesiyun, A. (2000). Microbial quality of water in rural communities of Trinidad. *Rev Panam Publica/Pan American J. Public Health*, *8*, 172-179.

Xu, H. Q. & Chen, B. Q. (2004). Remote sensing of the urban heat island and its changes in xiamen city of SE china. *Journal of Environmental Sciences (China)*, *16*(2), 276-281.

In: Flooding and Climate Change
Editors: D. D. Chadee, J. M. Sutherland and J. B. Agard
ISBN: 978-1-62948-317-7

Chapter 7

KNOWLEDGE, ATTITUDES AND PRACTICES OF COMMUNITIES TO FLOOD PREVENTION IN NORTH AND SOUTH TRINIDAD, WEST INDIES

Denny P. Singh[*] and Dave D. Chadee
Department of Life Sciences, The University of the West Indies,
St. Augustine, Trinidad, West Indies

ABSTRACT

In 2010 there were approximately 200 flooding episodes reported in Trinidad but little action had been taken by householders or local government authorities to introduce either new or old flood- adaptation strategies. This study was conducted to determine the knowledge, attitudes and practices of two communities, Barrackpore and St. Augustine, Trinidad, which have a history of major flooding events. A KAP questionnaire was developed using both categorical and quantitative questions of open, closed, and semi-closed styles. Questions covered demography and socioeconomics parameters, as well as, knowledge and understanding of the causes of flooding, and to elicit the communities' attitudes and practices for preventing or coping with flooding episodes. A total of 359 questionnaires were administered in the two flood-prone areas of Trinidad, 45% (163) in Barrackpore and 55% (196) in St. Augustine. Ninety-eight percent of the respondents from Barrackpore and 55% from St. Augustine knew that where they lived was flood-prone. They identified the main causes of flooding as poor drainage and proximity to rivers, and they identified the time at which flooding was likely to occurr, but did not develop flood protection strategies, or a flood recovery plan. Special attention should be paid to improving the drainage systems, implementing an early warning system for flood-prone areas, educating householders on flood prevention, developing community participation programs in collaboration with government to prevent dumping of garbage /debris in water courses, and providing general health and mental health care in both the short and long term.

[*] Corresponding author: E-mail: Dave.chadee@sta.uwi.edu.

1. BACKGROUND

In the 21st century Knowledge, Attitudes and Practice (KAP) surveys are accepted as suitable tools for collecting data on community perspectives and human behaviour - data required for primary health care at local, regional, national and international levels (Manderson and Aaby, 1992). This approach was established in the 1950s to deal with family planning issues, but has now expanded its range to include population studies (Cleland, 1973), and has evolved as a tool which can lead to in-depth understanding of socio-cultural and economic aspects of public health. Hence, KAP surveys have developed and established appropriate methodologies which can investigate health behaviours (Hausemann-Muela, 2003) and provide insights into the disease-transmission dynamics of dengue (Rosenbaun et al., 1995), malaria (Hlongwana et al., 2009), leptospirosis (Mohan and Chadee, 2011), TB (Shetty et al., 2004), HIV (Meena et al., 2013), as well as many aspects of medical anthropology (Launiala, 2009).

KAP studies are qualitative studies designed to better understand health -related knowledge, attitudes and practices and to describe the major challenges encountered by householders/communities. In this chapter the knowledge, attitudes or behaviour of communities living in high-risk -flood-prone areas in Trinidad were assessed, and their potential role in flood prevention evaluated.

2. INTRODUCTION

Globally flooding accounts for over 40% of all natural disaster s and causes approximately half of all disaster -related deaths. The estimated number of deaths due to flooding from 1900-2006 was 84, 849 (Ohl and Tapsell, 2000, Rodriguez-Llanes et al., 2011), with drowning being the major cause of death (Noji, 1991, Abuaku et al., 2009). Previously flooding episodes were generally considered to be restricted to developing countries in the tropics where rainfall due to monsoons and hurricanes can overwhelm drainage systems. However, over the last few decades anthropogenic factors and climate -change forces have led to extreme events resulting in flooding becoming widespread in both temperate countries (England and Wales, USA, Australia) and tropical climes (whether developed or developing) due to changing rainfall patterns from low intensity rainfall extended over long periods, to high intensity rainfall over short periods, poor infrastructure and little or no maintenance of drainage systems (Tunstall et al., 2006, Shea, 2007, Kovats and Akhtar, 2008).

In the developed world infectious disease outbreaks usually follow flooding episodes when the volume of water results in sewage backing-up and flowing into potable water systems, in damage to sanitation, sewage and water-treatment plants, disruption of refrigeration and cooking facilities due to electrical power cuts, and to people crowding into temporary make-shift shelters (Shea, 2007; Webster et al., 2005). In the developed world the impacts of disease outbreaks are generally mild and controlled, with minimum difficulty, whereas in developing countries where emergency planning may be poor or lacking, disease outbreaks can be deadly, extending over protracted periods, causing destruction of homes, disruption of power and water supplies, and delays in restoring these essential services (Shea, 2007; Webster et al., 2005; McMichael et al., 2001).

Epidemiological studies in low income rural and urban areas show evidence of increased risk of diseases associated with flooding such as diarrhea (Biswas et al., 1999), cholera (Walton and Ivers, 2011), respiratory infections (Shea, 2007), leptospirosis (Mohan and Chadee, 2011) and post-traumatic stress disorder (Ohl and Tapsell, 2000).

Within the Caribbean region flooding episodes are common, and usually (but see chapter 12) coincide with the rainy and hurricane seasons, and also occur in low-lying coastal areas where tidal fluctuations and sea-level rise impacts are being observed (Pulwarty et al., 2010). In most Small Island Developing States (SIDS) water supplies and drainage infrastructure are just adequate in cities, but in rural communities, water supplies are grossly inadequate or poor. Very few new development projects have been conducted to keep pace with increases in population and their demand for potable water during the first decade of the 21st century. In addition, climate -change forces have affected regional water supplies, with extreme events such as drought and flooding, leading to increased flooding disasters in cities located in coastal areas which are already also being affected by sea-level rise (IPCC 2007).

In 2010 there were approximately 200 flooding episodes reported in Trinidad but little action had been taken by householders or local government authorities to introduce either new or old adaptation strategies such as modifying the drainage infrastructure, developing capital works projects to improve drainage systems, or removing household items and environmental debris which block major water -courses. Rather a short-term strategy involving clean-up campaigns is usually adopted after flooding episodes to restore homes and communities and provide funding to householders to replace furnishings lost due to flood water damage. In the past no attempts were made to assess the knowledge, attitudes or behaviour of communities living in these high-risk -flood-prone areas in Trinidad, to evaluate their potential role in flood prevention. This study was conducted to determine the knowledge, attitudes and practices of two communities, one in north Trinidad and the other in south Trinidad both of which have histories of major flooding events. The information provided by this study will help in better understanding the action and attitudes of the population towards flooding for future planning and development.

3. Methodology

3.1. Study Area and Sample Size

The study was conducted during the period January to May 2011 which is considered part of the dry season on the island of Trinidad, which has an area of 4768km^2 and a population of 1,241,296 persons (CSO 2012). Numerous townships in Trinidad are prone to flooding during the rainy season (May to November) due to the geography of the land, proximity to rivers, and to anthropogenic factors (Figure 7.1). The Daniel formula was used to calculate the sample size of 359 persons. Variables were based on a 95% confidence level, and a prevalence of 5% was used with a precision of 4%.

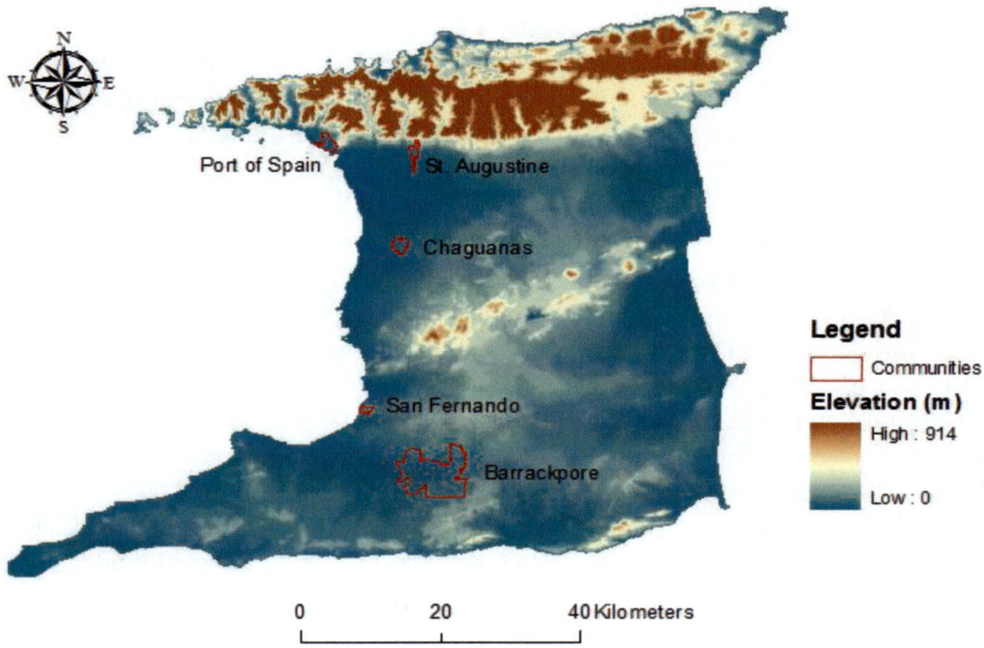

Figure 7.1. Showing the Barrackpore and St. Augustine study areas and the elevation of the land in Trinidad, West Indies. (Courtesy R. Mahabir, see chapter8).

3.2. Surveys (Data Collection)

A KAP questionnaire was developed by the research team and pre-tested using 20% of a sample population not used in the study. All required amendments were made to the questionnaire and 400 copies printed. The KAP instrument consisted of both categorical and quantitative questions of open, closed and semi-closed styles. Questions covered demography and socioeconomics parameters, as well as, knowledge and understanding of the causes of flooding, and to elicit their attitudes and practices to prevent or cope with flooding episodes.

The study used a cross-sectional design, and the selection of householders was conducted using a convenience sampling method as the surveys were conducted during the daytime. Only one person per household, and only persons over 18 years of age, were interviewed. All interviews were conducted by the authors from the University of the West Indies, St. Augustine, Trinidad, West Indies.

3.3. Data Analysis

All the KAP questionnaires were collected and the data encoded and organized into a database using Statistical Package for Social Sciences (SPSS) 18 (SPSS Inc. Chicago, IL, USA). The X^2 test was used to compare proportions and results were considered significant at the 5% level of significance (P<0.05). The results are presented in Tables and Figures.

4. Results

4.1. Demographics

From a total of 359 questionnaires administered in the two flood-prone areas of Trinidad, 45% (163) were conducted in Barrackpore and 55% (196) in St. Augustine. The difference in the sample numbers was due to the difference in population size and the size of the geographical area covered (Table 7.1).

Table 7.1. Showing the population size of the St. Augustine and Barrackpore study sites from Trinidad, West Indies (2011)

Flood Areas		
Regional Corporations	**Population**	**No. surveyed**
Tunapuna/Piarco (St. Augustine)	203,975	196
Debe/Penal (Barrackpore)	83,609	163
Total	287,584	359

Table 7.2. Demographics of the survey populations in Trinidad (2011)

Demographics	Barrackpore	St. Augustine	Total
Demographic Characteristic	**No. (%)**	**No. (%)**	**No. (%)**
Sex			
Male	77 (47%)	84 (42%)	161 (44.8)
Female	86 (53%)	112 (58%)	198 (55.2)
Age Group			
15-20	1 (0.6%)	3 (1.5%)	4 (1.1)
21-30	39 (23.9%)	50 (25.5%)	89 (24.7)
31-40	68 (41.7%)	84 (42.8%)	152 (42.3)
41-50	35 (21.4%)	43 (21.9%)	78 (27.7)
51-60	8 (4.9%)	6 (3.0%)	14 (3.8)
61 and over	12 (7.5%)	10 (5.1%)	22 (6.1)
Ethnicity			
Afro-Trinidadian	1 (0.6%)	1 (0.7%)	2 (1.1)
Indo-Trinidadian	152 (93.9%)	174 (89.8%)	326 (90.8)
European Decent	0 (0%)	0 (0%)	0 (0)
Chinese	0 (0%)	0 (0%)	0 (0)
Mixed	10 (5.5%)	17 (9.5%)	27 (8.1)
Education			
Primary	82 (50.0%)	7 (3.5%)	89 (24.7)
Secondary	71 (43.0%)	139 (70.9%)	210 (58.5)
Tertiary	10 (7.0%)	50 (25.6%)	60 (16.8)

At both study sites more females (>55%) were interviewed than males with the majority of persons belonging to the 21 to 60 age group (95%) in St. Augustine and to the 21 to 50 age group (90%) in Barrackpore (Table 7.2). The majority of the people who experienced flooding episodes were of East Indian heritage in St. Augustine (80%) and in Barrackpore (90%). The majority of the households 58.5% had secondary school education (Table 2), with significantly (P>0.03) larger number in St. Augustine (70%) than in Barrackpore (43%) and significantly larger number (P>0.02) having only primary school education in Barrackpore (50%) than in St. Augustine (3.5%).

Table 7.3. Knowledge and Practice of Interviewees on flooding in Barrackpore and St Augustine

	Barrackpore		St. Augustine		Total
Questionnaire Variable	**Total**	**No. (%)**	**Total**	**No. (%)**	**No. (%)**
Were you aware of the flooding before you took residence?	**163**		**196**		359
Yes		159 (98.23)		115 (59.12)	279 (77.7)
No		2 (1.80)		81(38.99)	83 (32.3)
Did you know that your neighbors experience flooding?	**163**		**196**		359
Yes		119 (72.33)		141 (72.33)	260 (72.4)
No		41(25.7 9)		49 (25.79)	90 (25.0)
Other		3(1.88)		6 (1.887)	9 (2.6)
What do you think is the main cause of flooding?	**163**		**196**		359
Poor drainage		147 (90.0)		105 (53.5)	253 (53.5%)
Proximity to river		16 (8.9)		45 (22.9)	61 (16.9%)
Blocked drains		0		25 (12.7)	25 (6.9%)
Other		0		21 (10.9)	21 (5.8%)
How many times have you experienced flooding that disrupted your daily lives?	**163**		**196**		359
Every time		120 (74.19)		149 (76.04)	269 (74.9)
Never		39 (24.19)		24 (12.50)	63 (17.5)
1-2 times		4 (1.62)		23 (11.40)	27 (7.6)
>3		0		0	0

	Barrackpore		St. Augustine		Total
Questionnaire Variable	**Total**	**No. (%)**	**Total**	**No. (%)**	**No. (%)**
Do you have any flood protection methods in place?	**163**		**196**		359
Sump pump		1 (0.62)		1(0.62)	2(0.5)
Backup power supply		0		0	0
Waterproof walls		13 (8.17)		19 (10.06)	32 (8.9)
Other		10 (6.28)		6 (3.145)	16 (4.4)
No methods		41 (25.79)		49 (25.79)	90 (25.1)
Regraded yard		98 (59.12)		121 (62.83)	219 (61.1)
Do you have a recovery plan?	**163**		**196**		359
Yes		0		0	0
No		163 (100)		196 (100)	359 (100)

g (X^2 =59.3; df 2; P>0.04). However, there was no significant difference in age of participants, or whether they had experienced flooding episodes (X^2 4.67; df 5; P<0.55).

4.2. Knowledge and Awareness of Flooding

Table 7.3 shows the responses of participants with respect to flooding ; their general knowledge and experience of flooding episodes in north and south Trinidad. A total of 77% (279) of the respondents knew that the areas they lived in were flood prone, 98% in Barrackpore and 59% in St. Augustine. In addition, over 72.3% of the participants in both Barrackpore and St. Augustine knew their neighbors had experienced flooding. They identified the main causes of flooding as poor drainage, and proximity to rivers. These accounted for 90% and 8.9% of responses in Barrackpore, and 53.5% and 22.9% of responses in St. Augustine (Table 7.3). There were significant differences between the communities in Barrackpore and St. Augustine in levels of education (X^2 = 10.43; df 2; P>0.02), as well as experience of flooding

4.3. Attitudes and Practices towards Flooding

Although the residents in Barrackpore (74.19%) and in St. Augustine (76.04%) reported disruption to their lives every time flooding occurred, they seemed not to have developed flood protection methods, and no one had developed a flood recovery plan (Tables 7.3 & 7.4). Table 7.4 clearly shows that the residents in both Barrackpore (58%) and St. Augustine (52%) were concerned about their welfare and frequently thought and spoke about flooding. However, these residents did not consider that flooding would affect their daily lives, but 70.1% and 69.0% in Barrackpore and St. Augustine, respectively felt flooding was a direct threat to their lives and felt that there was nothing they could do to prevent flooding, (69.9% in Barrackpore and 61.2% in St. Augustine).

The residents identified the months of major flooding in Barrackpore and St. Augustine and identified two major peaks in flooding, one from May to June and the other from August to November (Figure 7.2). In both study sites the residents agreed that something could be done about flooding, and also agreed that flooding could be a threat in the future (76% vs 75%), but in St. Augustine a significant (P>0.04) number of residents felt a damaging flood could occur in the future (Table 7.5). Both communities agreed that flood prevention and preparation strategies would alleviate flooding (90.4% and 81.3%), but were uncertain about the cost of flood preparations, and lacked the knowledge to prepare for flooding. The St. Augustine residents considered that they had the necessary skill to prevent flooding but required help or man power to prepare for flooding. Despite this knowledge, the residents' attitudes toward flood prevention seem similar, with 46% and 51% in Barrackpore and St. Augustine respectively being dependent on government help, and they both considered other issues more important to prepare for than flooding (66.6% and 58.3% respectively) (Table 7.5). In fact, when the residents of Barrackpore were asked about the most important problem facing their community 32.9% indicated poor drainage followed by flooding (27.3%), mosquitoes (21.7%), and rats (18.2%), whereas in St. Augustine the main problems were identified as flooding (24%), and mosquitoes (23%), followed by poor drainage (18%), rats (11.2%), flies (6.0%), burglary (4.8%) and poor roads (3.1%). The cost of flood damage in St. Augustine ranged from US $500.00 to US$3,500.00, while in Barrackpore the cost ranged from US$500.00 to US$1,000.00, however, none of the residents had insurance so they were dependent on government handouts and flood relief funding programs. In addition, the residents in both Barrackpore and St Augustine indicated that they received no flood warning, and were not satisfied with the flood emergency /relief measures provided.

Table 7.4. Assessment of the knowledge and attitudes of residents to the risk of flooding in Barrackpore and St. Augustine, Trinidad (2011)

Knowledge and attitude questions	Barrackpore (n= 163)	St. Augustine (n=196)	Total (n=359)
I think of flooding			
Not at all	20 (12.2%)	23 (11.7%)	43 (11.9%)
Sometimes	65 (39.8%)	43 (22.0%)	108 (30.0%)
A great deal	78 (48.0%)	130 (66.3%)	208 (58.1%)
I talk about flooding			
Not at all	37 (22.6%)	27 (13.7%)	64 (17.8%)
Sometimes	74 (45.5%)	35 (17.8%)	109 (30.3%)
A great deal	52 (31.9%)	134 (68.5%)	186 (52.9%)
I think a flood can pose a threat to my daily activities			
Not at all	111 (68.1%)	125 (63.8%)	236 (65.7%)
Sometimes	14 (8.5%)	52 (26.5%)	66 (18.3%)
A great deal	38 (23.4%)	19 (9.7%)	57 (16.0%)
I think a flood can pose a threat to personal safety			
Not at all	20 (12.2%)	25 (12.7%)	45 (12.5%)
Sometimes	29 (17.7%)	36 (18.3%)	65 (18.1%)
A great deal	114 (70.1%)	135 (69.0%)	249 (69.4%)

Table 7.5. Attitudes and practices of residents to flooding episodes in Barrackpore and St. Augustine, Trinidad, West Indies (2011)

	Responses				
Questionnaire Variable	**Strongly Agree No./(%)**	**Agree No./(%)**	**Neutral No/(%)**	**Disagree No/(%)**	**Strongly Disagree No/ (%)**
It isn't useful preparing for floods because they are too destructive					
Barrackpore	5 (3.1)	0 (0)	3 (1.5)	21 (12.6)	134 (82.5)
St. Augustine	16 (8.3)	14 (7.2)	10 (5.2)	20 (10.4)	136 (69.1)
A flood is unlikely to occur during our lifetime					
Barrackpore	5 (3.2)	3 (1.5)	3 (1.5)	28 (17.5)	124 (76.3)
St. Augustine	4 (2.0)	14 (7.2)	18 (9.3)	13 (6.5)	148 (75.0)
Preparation is unnecessary because the local government will provide assistance					
Barrackpore	25 (15.8)	0 (0)	49 (30.2)	2 (1.5)	87 (52.5)
St. Augustine	10 (5.2)	16 (8.3)	73 (37.5)	18 (9.4)	79 (39.6)
A damaging flood can occur in the future					
Barrackpore	69 (41.0)	15 (9.8)	36 (22.0)	21 (13.1)	22 (14.1)
St. Augustine	115 (58.3)	42 (21.8)	16 (8.3)	8 (4.2)	15 (7.4)
Preparation and preventative methods should be taken to alleviate flooding					
Barrackpore	148 (90.4)	5 (3.1)	0 (0)	8 (4.8)	2 (1.7)
St. Augustine	160 (81.3)	20 (10.4)	16 (8.3)	0 (0)	0 (0)
There is too great a cost associated with flood preparedness					
Barrackpore	45 (26.9)	2 (1.5)	15 (9.5)	47 (28.6)	54 (33.5)
St. Augustine	43 (21.8)	24 (12.5)	43 (21.8)	31 (15.6)	55 (28.3)
You lack the skills required to prepare for flooding					
Barrackpore	2 (1.5)	8 (4.7)	68 (41.3)	62 (38.0)	23 (14.5)
St. Augustine	13 (6.5)	17 (8.8)	25 (12.6)	37 (18.9)	104 (53.2)
You lack the proper knowledge of flooding to prepare					
Barrackpore	5 (3.1)	10 (6.3)	84 (51.0)	49 (30.2)	15 (9.4)
St. Augustine	27 (14.5)	48 (22.9)	14 (7.2)	34 (17.7)	73 (37.7)
There are many other important things to prepare for other than floods					
Barrackpore	60 (36.5)	49 (30.1)	10 (6.3)	41 (25.3)	3 (1.6)
St. Augustine	65 (28.1)	59 (30.2)	6 (3.1)	14 (7.2)	61 (31.4)
You require added help or man power to prepare for flooding					
Barrackpore	21 (12.6)	21 (12.6)	91 (55.5)	18 (11.1)	23 (7.9)
St. Augustine	72 (36.7)	51 (26.0)	8 (4.1)	22 (11.4)	43 (21.8)

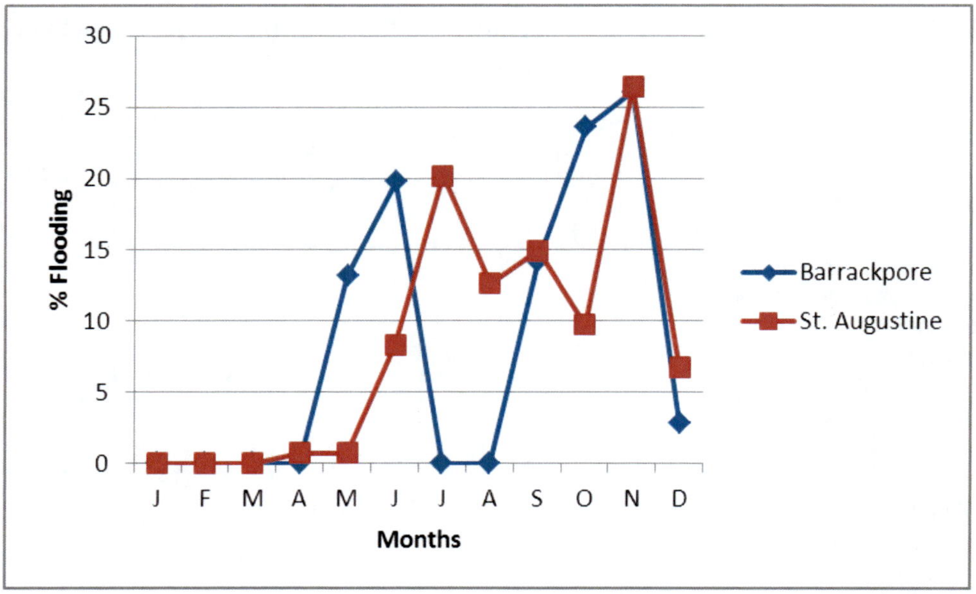

Figure 7.2. Showing the months of major flooding in Barrackpore and St. Augustine, Trinidad identified by householders.

5. Discussion

With the increase in the incidence of flooding worldwide it is surprising that this represents the first KAP study on the subject, especially as flooding is known to: cause death due to drowning, and injuries; and to increase infectious disease outbreaks; cause loss of property and destruction of agricultural crops ; and can traumatize affected populations (IPCC, 2007; Kovats and Akhtar, 2008). During the present study most residents surveyed knew that the areas in which they lived were flood-prone and they, and/or their neighbours, had experienced flooding episodes. The residents of both Barrackpore and St. Augustine not only knew of the risk of flooding in their neighbourhoods, but also identified the time of year during which major flooding occurred (Figure 7.2), that is, they identified two different peaks, May to June and September to November in Barrackpore, and June to July and September to November in St. Augustine but did little to prevent flood damage to their properties.

Since the residents of both Barrackpore and St. Augustine knew the main causes of flooding and have all experienced flood damage, it is important to determine why the residents continue to live in these areas. In the Barrackpore area the land belongs to the residents and is used to cultivate sugar cane, vegetables and rice, while in the St. Augustine area many of the residents are squatters or people without security of land tenure, but the location provides access to goods and services, and is in close proximity to major townships such as the city of Port of Spain (12 miles), the Borough of Chaguanas (8 miles) and the university town of St. Augustine (< 1 mile). However, the residents with security of tenure were unable to move due to the high cost of properties elsewhere and their present location provided close proximity to their jobs, access to health care, and excellent goods and services. Similar patterns of behaviour have been reported recently in South East Asia n countries including India, Pakistan, and Bangladesh, and also in Caribbean and Latin America n countries including Guyana, Suriname and Haiti where people balance the risks and benefits of living in flood-prone areas, and their ability to access low-cost housing, work and health care (Ahern et al., 2005; UN Human Settlement Programme, 2011).

The large population of people of East Indian heritage in Barrackpore and St. Augustine reflects historical land use practices which involved indentured labourers being brought to the sugar cane estates in Trinidad during the period 1843 to 1917. Many were provided housing, therefore generations have lived in these low lying areas for over 150 years, some with, and some without, security of tenure. The security of tenure issue may also prevent residents from spending money to develop flood prevention or adaptation strategies because they anticipate/hope to be relocated by government programs. In addition, insurance companies may have also labeled these areas as being high risk and refuse to write policies to cover flood damage. However, after every major flood episode the government provides assistance with cleaning of residences, and by replacing furniture and appliances damaged by flood water. It is instructive to review the responses of the residents from both Barrackpore and St. Augustine when asked about flood prevention and government assistance: the majority of the residents remained neutral not agreeing or disagreeing with the dependency on government for flood relief in St Augustine; but 52.5% strongly disagreed in Barrackpore. The differences in attitudes may also be reflected in the differences in the cost of flood damage – that in Barrackpore being between US$500.00 to US$1,000.00 while that in St Augustine was between US$500.00 to US$3,500.00. It is possible that the time required for post-flood recovery may be shorter in Barrackpore because of community assistance, than in St. Augustine where most flood victims depend on government assistance and work by local authorities. Further studies on the cost of flooding and the dependency of communities on flood relief are recommend in order to develop flood-adaptation strategies to prevent future flooding episodes.

The lack of knowledge of how to prepare for floods has been highlighted by residents of both Barrackpore and St. Augustine because they are all concerned that in the future more severe floods may occur. Studies on climate variability and climate change suggest that health education and health promotion, as well as flood prevention and adaptation solutions should be developed. These solutions will empower communities by providing the necessary technical and logistic backstopping required to prevent the loss of life and spread of disease associated with flooding, as well as the psychological impacts associated with these extreme events (Ohl and Tapsell, 2000. Kovats and Akhtar, 2008)

In addition, the attitude of the interviewees towards flood prevention suggests that 61% had become fatalistic with significant (X^2 9.26; d.f. 2; $P<0.03$) numbers thinking that flooding of their homes cannot be prevented, that is, 69.3% and 54.6% in Barrackpore and St. Augustine, respectively. The fact that the householders had no recovery plan and did not take measures to prevent flooding of their homes suggests that some complex behavioural disorder has become established in the population over time. Although, this study was not designed to unearth mental health effects associated with flooding these factors are certainly interesting and suggest further studies would be more revealing. Numerous studies have identified mental health symptoms in flood victims (Mason et al., 2010, Murray et al., 2011), and in England and Wales, Paranjothy et al. (2011) reported psychological distress, anxiety, depression and post- traumatic stress syndrome as features common to individuals exposed to flooding when compared with groups not exposed to flooding episodes.

Many of the residents from both Barrackpore and St Augustine agreed that there were more important things to prepare for than flooding (Table 7.5) but were uncertain about the cost associated with flood adaptation strategies. More than 62% of the respondents from St. Augustine agreed they needed help or man power to prepare for flooding while in Barrackpore the respondents remained neutral by not agreeing or disagreeing. These results suggest that the residents wanted adaptation strategies for flood prevention but were not prepared to conduct them without outside assistance. This attitude is possibly the result of continued or long term effects of stress associated with flooding events, loss of property, disruption of lives, no security of tenure, and anxiety, thus making people depressed and dependent on others. Similar findings have been reported regarding flood victims in China and Pakistan (Dar et al., 2011), and in England and Wales (Mason et al., 2010).

CONCLUSION

It is clear from this study that special attention should be paid to improving the drainage systems, implementing an early warning system for flood-prone areas, educating householders on flood prevention, developing community participation programs, and implementing appropriate legislation to prevent dumping of debris in water courses, as well as providing general health and mental health care in both the short and the long term. Studies are currently being conducted to address some of these issues.

REFERENCES

Ahern, M., Kovats, R. S., Wilkinson, P., Few, R. & Matthies, F. (2005). Global health impacts of floods : Epidemiologic evidence. *Epidemiologic Reviews*, *27*, 36-46.

Abuaku, B. K., Zhou, I., Li, X., Li, S., Li, A., Yang, T. & Tan, H. (2009). Morbidity and mortality among populations suffering flood in Hunan, China. The role of socioeconomic status. *Journal of Flood Risk Management*, *2*, 222-228.

Biswas, R., Pal, D. & Mukhopadhyay, S. P. (1999). A community based study on health impact of flood in a vulnerable district of West Bengal, *43*, 89-90.

Dar, O. A., Khan, M. S. & Murray, V. (2011). Conducting rapid health needs assessment in the cluster era: experience from the Pakistan flood. *Pre-hospital Disaster Medicine*, *26*, 212-216.

CSO. (2012). *Population and Housing Census Register*. Central Statistical Office, Port of Spain : Ministry of Planning and Development: 2012

IPCC. (2007). Climate Change 2007: Impacts, Adaptation and Vulnerability. Contribution of Working Group II to the Fourth Assessment Report of the Intergovernmental Panel on Climate Change [Parry, M.L., O.F. Canziani, J.P. Palutikof, P.J. van der Linden, and C.E. Hanson (eds.)]. Cambridge University Press, Cambridge, UK, 976 pp.

Kovats, S. & Akhtar, R. (2008). Climate, climate change and human health in Asian cities. *Environment and Urbanization*, *20*, 165-175.

Mason, V., Andrews, H. & Upton, D. (2010). The psychological impacts of exposure to floods. *Psychology, Health and Medicine*, *15*, 61-73.

McMichael, A. J., Haines, A. & Kovats, R. S. (2001). Methodological Issues in Assessing the Effects of Climate Change on Health In: Green E, Maynard R (eds). *Health Effects of Climate Change in the UK.* London.

Mohan, A. R. M. & Chadee, D. D. (2011). Knowledge, attitudes and practices of Trinidadian households regarding leptospirosis and related matters. *International Health*, *3*, 131-137.

Murray, D., Caldin, H., Amlot, R., Stanke, C., Lock, S., Rowlatt, H. & Williams, R. (2011). *The effects of flooding on mental health.* Health Protection Agency p. 127.

Noji, E. (1997). *The Public Health Impacts of Disasters*. Oxford University Press, New York, NY.

Ohl, C. A. & Tapsell, S. (2000). Flooding and human health. *British Medical Journal*, *321*, 1167-11

Paranjothy, S., Gallacher, J., Amlôt, R., Rubin, G. J., Page, L., Baxter, T., Wight, J., Kirrage, D., McNaught R. & Palmer, S. R. (2011). Psychosocial impact of the summer 2007 floods in England. *Public Health*, *11*,145.

Pulwarty, R. S, Nurse, L. & Trotz, U. O. (2010). Caribbean Islands in a changing climate. *Environment: Science and Policy for Sustainable Development, 52*, 16 – 27.

Shea, K. M. (2007). Global climate change and children 's health. *Pediatrics. 120*, e1359-1367.

Tunstall, S., Tapsell, S., Green, C., Floyd, P. & George, C. (2006). The health effects of flooding: social research results from England and Wales. *Journal of Water and Health*, *4*, 365-380.

United Nations Human Settlement Programme (2011). *Cities and Climate Change*, UN Habitat, London.

Walton, D. & Ivers, L. C. (2011). Responding to cholera in post-earthquake Haiti. *New England Journal of Medicine*, *364*, 3-5.

Webster, P. J., Holland, G. J., Curry, J. A. & Chang, H. R. (2005): Changes in tropical cyclone number, duration, and intensity in a warming environment. *Science*, *309*(5742), 1844-1846

In: Flooding and Climate Change
Editors: D. D. Chadee, J. M. Sutherland and J. B. Agard
ISBN: 978-1-62948-317-7

Chapter 8

APPLICATIONS OF REMOTE SENSING AND GIS TECHNOLOGIES IN FLOOD RISK MANAGEMENT

Raid Al-Tahir, Intesar Saeed and Ron Mahabir
The Centre for Caribbean Land and Environmental Appraisal Research (CLEAR)
The University of the West Indies, St Augustine, Trinidad and Tobago

ABSTRACT

Flooding is the most common of all major disasters that regularly affect populations and results in extensive damage to property, infrastructure, natural resources, and even to loss of life. To ensure better outcomes, planning and execution of flood management projects must utilize knowledge on a wide range of factors, most of which are of a spatial nature. Advances in geospatial technologies, specifically remote sensing and Geographic Information System s (GIS), have enabled the acquisition and analysis of data about the Earth's surface for flood mitigation projects in a faster, more efficient and more accurate manner.

Remote sensing and GIS have emerged as powerful tools to deal with various aspects of flood management in prevention, preparedness and relief management of flood disaster. GIS facilitates integration of spatial and non-spatial data such as rainfall and stream flows, river cross sections and profiles, and river basin characteristics, as well as other information such as historical flood maps, infrastructures, land use, and social and economic data. Such data set s are critical for the in-depth analysis and management of floods.

Remote sensing technologies have great potential in overcoming the information void in the Caribbean region. The observation, mapping, and representation of Earth's surface have provided effective and timely information for monitoring floods and their effect. The potential of new air- and space-borne imaging technologies for improving hazard evaluation and risk reduction is continually being explored. They are relatively inexpensive and have the ability to provide information on several parameters that are crucial to flood mapping and monitoring.

This chapter examines the use of the new imagery data as an up-to-date and affordable source of information for establishing the necessary information for flood management in the Caribbean region. The chapter also reviews the use of an integrated GIS approach to fulfil data and analysis needs for planning and emergency operations and how it can become the backbone of emergency management.

1. INTRODUCTION

Of the various geohazards that disrupt the natural and man -made environments, flooding is the most common and most costly. In fact, of all natural risks, floods pose the most widely occurring hazard to life today. A flood is commonly defined as any level of flow that exceeds the natural carrying capacity of a watercourse, overflows its banks, and inundates the adjacent low lying lands (OAS 1991). The United States Geological Survey defines a flood as a "relatively high water that overflows the natural or artificial banks of a stream or coastal area that submerges land not normally below water" (USGS 2007).

The leading cause of floods is a prolonged heavy rainfall that creates high runoff in rivers, or a build up of surface water in areas of low relief. On the other hand, intense rainfall over short periods from storms and hurricanes may produce rapid runoff and sudden but severe flash floods across river valleys, resulting in insurmountable losses to human life and property compared to any other type of flooding. Coastal areas are also prone to flooding when storms and large waves bring seawater onto the land (Few et al., 2004). Small Island Developing States are especially impacted since a great majority of the population live and work near coastal areas. Other conditions that may cause flooding are those events that occur less frequently but result in major consequences when they occur. These include events such as earthquakes, bursting of reservoirs, failure of dams, dikes and weir gates, and tsunamis.

Specific to developing countries, including the Caribbean region, the flooding problem is further intensified by inappropriate land use and environmental degradation in areas that are vulnerable to erosion, rapid water runoff, and slope failure. Poor land-use practices (e.g., slash and burn agriculture, quarrying, deforestation, and rapid urbanization and development) lead to heavy sedimentation in river channels, thus, reducing their carrying capacity. In parallel, the removal of vegetative cover leads to much shorter lag times between rainfall and the water reaching the waterways causing the already reduced channels to overflow, leading to massive floods (Ramlal and Baban 2008).

Further upset in flood events is also expected as a result of global climate change, leading to consequent increase in extreme climatic events such as severe dry spells, heavy rainstorms and tidal surges. It is expected that areas most vulnerable will continue to be further devastated (Few et al., 2004; Ramlal and Baban 2008). Such findings are supported by reports made available by the Intergovernmental Panel on Climate Change and related research from scientific bodies with similar interests worldwide.

According to the UN World Water Assessment Programme, floods accounted for 15% of all deaths related to natural disaster s, and approximately 66 million people suffered flood damage from 1973 to 1997 (WWAP 2006). Persons can be both directly and indirectly impacted by floods. Direct impacts include loss of lives, damage to property, disruption of transportation, communications, health and community service s, crop and livestock damage, and interruptions and losses in businesses. The indirect effect s of floods are far more damaging and include elements of society that reflect a more in-depth and personal assessment of flooding on the wellbeing of people (Badilla 2008).

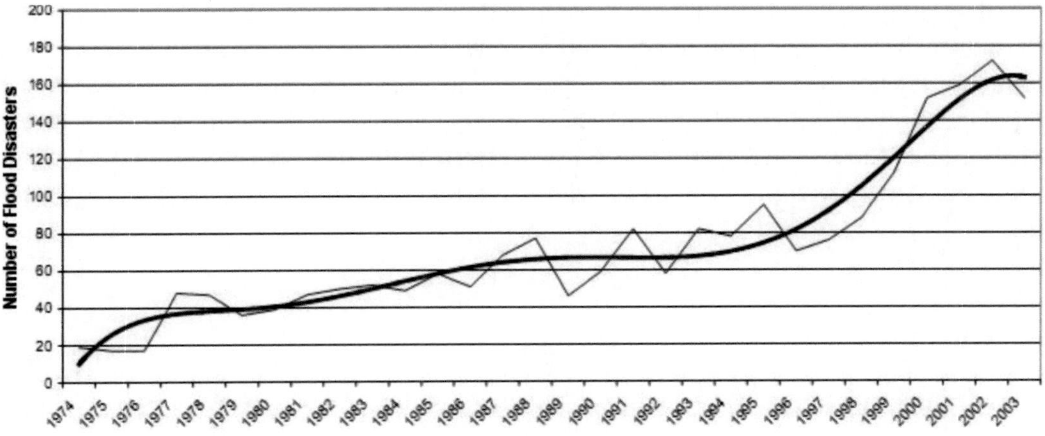

Figure 8.1. Total number of reported disasters caused by floods in the World: 1974 – 2003 (Hoyois and Guha-Sapir 2004).

Developing countries are disproportionately affected by disasters. Economic losses attributable to natural hazards in developing countries are about 5 times higher per unit of gross domestic product (GDP) than those of more affluent countries. Furthermore, developing countries accounted for over 50% of all disaster fatalities from 1992 to 2001, which translates into approximately 13 times more people die per reported disaster than in developed countries (Van Westen 2002; WWAP 2006). This problem is further exacerbated with knowledge that the magnitude and frequency of floods will continue to increase as indicated by the trend in Figure 8.1 (Badilla 2008; Hoyois and Guha-Sapir 2004).

Given the severe negative impacts of floods there is therefore an urgent need to introduce proper management measures to ensure that vulnerable areas are protected and flooding impact is minimized. Whether using structural, non-structural or mixed strategies, improving the capabilities to mitigate and respond to flood hazards would require better understanding and mapping of the hazards, as well as identifying the nature and extent of vulnerability of the areas under consideration. These activities prompt both the need and use of spatial data representative of the disaster and factors that can be used in assisting all phases of a comprehensive disaster management program.

This chapter is divided into 4 sections from this point on. Section 2 will review the different phases and activities of a disaster management program, and highlights the specific required spatial datasets within each phase. Section 3 will introduce, in general, the use of Geographic Information System s (GIS) and remote sensing techniques in flood management, while Section 4 will identify, in greater detail, specific studies and cases for the application of these technologies as they relate to the different phases of disaster management. Finally, section 5 provides conclusions.

2. Disaster Management Activities

Disaster management endeavours to reduce or avoid the potential losses from hazards, assure prompt and appropriate assistance to victims of disaster, and achieve rapid and effective recovery. Activities related to disaster management can be grouped into pre-event activities (mitigation and preparedness), actions during and immediately following an event (rescue and relief), and post-disaster activities (rehabilitation and reconstruction). These phases are related to each other and are typically represented as a cycle since the occurrence of a disaster event will eventually influence the way society is preparing for the next one (Mansor et al., 2004; OAS 1991; Van Westen 2002). The following sections will provide a synopsis of the activities in each phase with the objective of identifying the required spatial data.

2.1. Mitigation

Mitigation activities focus on eliminating or reducing the probability of a disaster occurrence and reducing the impact of unavoidable disaster. This phase involves essentially three kinds of studies (OAS 1991):

- *Natural Hazard Assessments* provide information on the probable location and severity of natural hazards and their likelihood of occurring within a specific time period in a given area;
- *Vulnerability Assessments* estimate the degree of loss or damage that would result from the occurrence of a natural hazard of given severity; and
- *Risk Assessments* integrate information from the analysis of an area's hazards and its vulnerability to them, into an estimate of the probability of expected loss from a hazardous event.

The effectiveness of disaster mitigation depends on the availability of information related to the hazards and the various sectors affected by them. As such, data collection and analysis is required to identify and evaluate appropriate measures. Hazard assessment studies rely heavily on geologic, geomorphic, and soil maps; climate and hydrological data; topographic maps, aerial photographs, and satellite imagery; as well as historical information on previous occurrences of the hazard. Vulnerability assessments, on the other hand, need to incorporate and analyse information on human population, capital facilities and resources, and economic activities (OAS 1991).

2.2. Preparedness

The goal of disaster preparedness is to develop plans that minimize the loss of life and damage to property during a natural event. Disaster preparedness promotes the development of a system for monitoring known hazards, a warning system, emergency and evacuation plans, and emergency routes. Additionally, this phase includes a number of efforts aimed at increasing the amount of information disseminated to the public and at promoting cooperation between the public and the authorities in case of an emergency (Johnson 2000; OAS 1991).

2.3. Rescue and Relief

Following the occurrence of the disaster, activities in this phase are designed to provide emergency assistance for victims that include search and rescue, emergency shelter, medical care, and mass feeding. They also seek to stabilize the situation and reduce the probability of secondary damage (Johnson 2000). The success of these activities is heavily dependent on real time information with respect to service routes and up-to-date information on the location and resource availability at the various critical facilities.

2.4. Post-Disaster Rehabilitation and Reconstruction

Post-disaster activities are necessary to return all systems to a state of normalcy or at least to a state that is better than that immediately following the occurrence of a disaster. These activities have a short-term goal of restoring vital life support systems to minimum operating standards. Additionally, this phase includes long-term rehabilitation activities to re-establish the functions of public services, business, and commerce, to repair housing and other structures, and to return production facilities to operation (Johnson 2000; OAS 1991).

Investigations at the post-disaster phase will re-examine the qualitative and quantitative aspects of natural hazards, often improving on information produced by modelling and conjecture, by indicating areas where development should be extremely limited or should not take place (OAS 1991).

3. Role of Geospatial Technologies in Flood Management

An effective flood management program, and all its phases, depends on a wide range of parameters and large volumes of accurate, relevant, and timely geoinformation. Besides, geographical data and tools are an absolute necessity if scientific understanding is to be translated into effective evidence -based and place-based policy (Murphy et al., 2010). As indicated in the previous section, the required physical information includes topography and terrain, soil types, watershed /catchments, land cover and forestry, and the intensity of the triggering factors. Ultimately, flood risk management requires socio-economic data (housing location, valuation data, demographic structure, census information) as well as land use information, administrative boundaries, development pressure, and environmental constraints.

However, there is a severe shortage of reliable and compatible data set s generally in the whole Caribbean region. Information needed for accurate planning is often outdated, non-existent, or very expensive and time consuming to collect (Al-Tahir et al., 2006). Without such information, the investigation of flood susceptibility, and the formation of proper national planning policies in many Caribbean island states will be both difficult and error prone.

Modern geospatial technologies, especially in the disciplines of photogrammetry, remote sensing, and spatial information science, can assist in crisis management through observation, mapping, and analysing relevant information in each of the critical disaster management phases (Altan and Kemper 2010). Geospatial data and tools have a key role in the saving of lives, the limitation of damage, and the reduction in the costs to society of dealing with emergencies. Recent improvement in computer software and hardware have allowed remote sensing and GIS to provide the way forward in collecting and managing relevant data set s, and in developing management scenarios to evaluate mitigation strategies (Ehlers 2004).

3.1. The Role of GIS

Flooding is a dynamic process and exhibits spatial characteristics in the sense that floods occur at a particular location at which various contributing factors for the event exist. For example, typical flood conditions include areas gently sloping or basins, heavy rainfall, saturated soil and blocked channels (Alkema 2004). The primary function of GIS is to store and link multiple digital databases and many different themed layers of information, graphically display that information as maps, and to examine how the layers interrelate. These layers, such as those depicted in Figure 8.2, would carry the key elements needed for flood hazard management (GAO 2004). This information may come from many sources, including remote sensing, GPS, censuses, soil samples, stream gage systems, and weather stations. Additionally, historical data help in identifying where flood may occur since areas most commonly flooded are expected to flood again.

Having acquired and stored the required data, GIS based analysis is then conducted to determine the amount of influence of each of the factors and their impacts in different scenarios. GIS provides an enabling technology by means of presenting a feasible solution, providing tools and methods that support all phases of flood management. Specifically, GIS is used for locating critical and vulnerable assets, planning, mitigation activities, assisting in response, and aiding in recovery management (Khanna et al., 2006). Table 1 lists various common applications of GIS in hazard management in general.

Table 8.1. *Common* applications of GIS in hazard management (Marfai 2003)

Function	Potential Applications
Data display	• Aid in the analysis of spatial distribution of socio-economic infrastructure and natural hazard phenomena. • Use of thematic maps to enhance reports and/or presentations. • Link with other databases for more specific information.
Land information storage and retrieval	Filing, maintaining, and updating land related data (e.g., land ownership, records of previous natural events, permissible uses, etc.)
Zone and district management	• Maintain and update district maps, such as zoning of floodplain maps. • Determine and enforce adequate land-use regulation and building codes
Site selection	Identification of potential sites for particular uses
Hazard impact assessment	Identification of geographically determined hazard impacts
Development/land suitability modelling	Analysis of suitability of particular parcels for development.

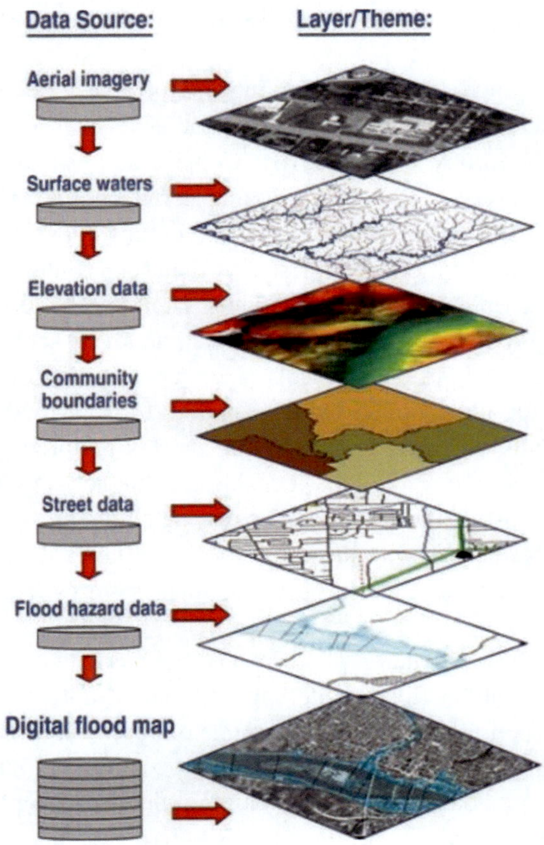

Figure 8.2. GIS Layers for Flood Mapping (GAO 2004).

By overlaying or intersecting different spatial layers, flood-prone areas can be identified, thereby enabling more effective mitigation and response to flooding events. GIS is also used to present flood-related information visually in 2D or 3D such that it might be more intuitively understood, particularly by a lay audience. GIS-generated maps provide an effective means of informing the public or news media on evacuation routes, on the extent and impact of a flood when it occurs, and on the government's response efforts (Shanley et al., 2006). This information can be disseminated with the use of web technologies, creating interactive maps that can be queried and updated in real time, and linked to additional information.

3.2. The Role of Remote Sensing

Remote sensing is another advancing technology that has benefited flood modelling and flood risk assessment in providing a cost effective means of acquiring accurate spatial and temporal data (Khanna et al., 2006). Remote sensing of the environment involves the measurement of electromagnetic radiation reflected from or emitted by the Earth's surface and relating these measurements to the types of land cover and habitat in the area being observed by the sensor (Al-Tahir et al., 2006). Observing Earth's surface using sensors on board various space and air platforms has evolved into a multinational enterprise producing a large amount of digital data on various physical parameters that characterize the earth system (Murphy et al., 2010).

Remote sensing continues to expand making use of new kinds of sensors that exploit different parts of the electromagnetic spectrum, and achieve finer levels of spatial, temporal, or spectral resolution (Murphy et al., 2010). Considerable advances in remote sensing technology have occurred both in acquiring digital aerial photography and high resolution satellite data. Parallel to this, new techniques have been developed for improved processing and extraction of spatial information from these new data set s (Ehlers 2004).

The distinction between aerial photography and satellite remote sensing is based on the fact that each of them provides some capabilities that cannot be achieved by the other. Such comparative capabilities include ground coverage, repeatability of observations, spectral ranges, and geometry for three dimensional mapping (Li 1998). Aerial photo data are of immense value, especially when long term spatial data is required, such as in the case of hydrology and flooding (Rango et al., 2008) considering that aerial photography extends back into the mid 1940s in most of the Caribbean states. Satellites on the other hand have been around for a much shorter period, with the early images of the earth's surface collected in the 1970s. The principal advantages in using satellite images stems from their higher spectral resolution.

Information from earth observation platforms, combined with other relevant data in GIS, have triggered advancements in flood modelling and demonstrated their utility in providing historical, recent, and repetitive data for a wide range of applications in disaster management. Earth observation techniques are used for mapping geomorphologic elements, historical events and sequential inundation phases, including duration, depth of inundation, and direction of current *to* indicate which areas are potentially vulnerable (Altan and Kemper 2010; Van Westen 2002). Moreover, images of an area can be analyzed together with various land cover input parameters, essential for use and development of hydrological modelling for risk analysis.

High resolution images offer great opportunities to identify individual structures and natural features as well as their functions that are important for the assessment of their vulnerability and value. This is a very critical aspect in monitoring the increasing risks and impacts of floods, especially for urban areas that experience fast and uncontrolled expansion into hazardous areas (Alkema 2004). Because of its availability and temporal resolution, high resolution imagery provides a critical tool for monitoring floods before, during and after the event, supporting a much more in depth analysis of the hazard and hydrological models.

Digital elevation model (DEM) (or alternatively, digital terrain model; DTM) is essential for flood management and hydrological modelling. DEM is used for deriving critical information sets such as possible sinks, flow direction, and rate of flow of floods. Such data is yet another data set that can only be provided effectively by remote sensing. Aerial surveys using airborne cameras and/or airborne LIDAR sensors are able to deliver a high density digital terrain model. DEM can also be extracted from space- borne sources such as satellite stereo images of SPOT, or radar data such as those collected by the Shuttle Radar Terrain Measurement (SRTM) data. Using DEM in combination with other data in a GIS analysis basin boundary, drainage network, area, and elevation can be extracted, and risk estimations can be carried out (Badilla 2008). The resolution of the DEM is critical since this will dictate the smallest structural features (e.g., rivers) that can be reconstructed (Alkema 2004).

4. Applications of Integrated Geospatial Technologies in Flood Risk Management

The following sections will review some specific cases where remote sensing and GIS have provided the tools for data collection and analysis in support of pre-disaster preparedness programs, in-disaster response activities, and post-disaster reconstruction.

4.1. Disaster Mitigation and Preparedness

The most promising application of remote sensing is its use for risk analysis as a function of hazard and vulnerability assessments. The mapping of areas affected by floods is essential for both planning and the assessment of incurred damages as a result of the disaster. Achieving this will require mapping flooded areas before and after flood waters recede. Satellite and aerial remote sensing provides a suitable alternative to traditional field surveying techniques for producing flood inundation maps. Besides, they offer the possibility of validating the output of developed models by comparing the observed extent of the flood with the modelled prediction (Alkema 2004; Wilson and Atkinson 2003).

Due to its advantageous synoptic view of the hazard, geospatial imagery can be added into the GIS as a layer after being georeferenced. The extent of flooding is then delineated and further superimposed onto additional spatial variables for analysis. Also, images representing the extent of flooding can be collected at different time intervals in order to monitor both the progression and recession of the flood inundation within a short period of time which can be used for planning and organizing mitigation or response operations effectively (Khanna et al., 2006).

Flood susceptibility mapping and risk area delineation is another task that can benefit from an integrated use of GIS and remote sensing. In a study carried out in Malaysia, a GIS and remote sensing -based flood susceptibility model was prepared using the known extent of flooded areas (Pradhan 2009). Historical flooded areas extracted from RADARSAT images, and terrain characteristics extracted from the DEM (e.g., slope, aspect, flow direction), soil, soil texture, land cover, and precipitation information were updated to enable the quantification of flood associated attributes. A flood susceptibility map was then produced using a logistic regression model that assessed all these factors.

Most studies typically assess flood risks purely in terms of economic damage while neglecting the social and economic impacts, leaving the map to assess the risk.

One may face a challenging task when incorporating several factors in identifying both potential areas to be flooded and the associated level of risk should the hazard be realized. This becomes an issue of concern since different factors exert different levels of influence based on present situation and their interactions within the environment. In this event weights must be given to the different criteria for indicating the level of impact that each risk criterion has on flooding. In studies by Meyer et al. (2009) and Wallingford (2007), a multi-criteria approach towards flood-risk management is given. GIS provides an optimum environment for providing different possibilities for multi-criteria analysis of spatial problems such as floods.

Hydrological models are used as another approach in flood hazard mapping in parallel with the use of historical events as the main indicator. A study by Kafle et al. (2006)

integrated flood simulation modelling with remote sensing and GIS data for flood hazard mapping in the Baghmati River, Nepal. Dry season satellite images were used to update the existing land use map in the flood affected areas. Following this, a hydrologic model was developed in combination with a digital elevation model, and was used for delineating the inundation area extent, and estimating the flood depths in areas where images capturing the peak flood events were not available. Consequently, flood hazard maps were prepared for a 50-year flood followed by designing a flood-hazard ranking system by combining flood extent, flood depth, land-use types and population. A specific large segment of the river bank was subsequently identified as the main spilling reach for structural countermeasures (Kafle et al., 2006).

It is evident from all previous discussions that digital elevation model (DEM) plays a very essential role in hydraulic models and risk analysis. As such, several studies have focused on the varying means for acquiring elevation data, and the impact of their resolution and accuracy on the analysis. Such a study was conducted by Wilson and Atkinson (2003) for the use of LiDAR, stereo photogrammetry, and ERS interferometric SAR (InSAR) elevation data for flood inundation modelling. Each dataset was used to predict flood inundation for a 60 day event during a series of major flooding of the Avon River, England, that occurred in autumn 2000. Predictions using each elevation dataset were then compared to aerial photography of the flooding acquired close to the flood peak. The most accurate prediction of inundation was obtained using the LiDAR data (after the removal of above surface features), followed by photogrammetric data (Wilson and Atkinson 2003).

In another study, concerns over the suitability of using a coarse resolution of DEM (50 metre) to reconstruct small structural features of small rivers, especially those with narrow banks like Onga River on Kyushu Island in Japan were investigated (Mori 2007). To achieve the goals for this study, a 2-meter meshed DEM was generated using the airborne laser system with a vertical resolution of 0.1 metre. For estimating flooded areas around the Onga River, 3D topographic maps and 1-meter contour lines were created using the DEM data in the GIS system. The study concluded that the higher resolution DEM was the key for providing precise water levels and, hence, more accurate estimations of the flooded area.

4.2. During and Post-Disaster Applications

During the event of flooding, the integrated use of remote sensing and GIS has been effectively applied in issuing warnings (such as cyclone tracking), appraising the extent of damage, and providing alternative evacuation routes during rescue and relief efforts. Developed evacuation plans need to be rapidly updated in response to new images and information over the affected population as well as the damaged routes and shelters. Figure 8.3 depicts such a GIS and remote sensing approach to dynamic response to a disaster. Such an approach, though, depends on a rich source of accurate and up-to-date GIS and auxiliary data, and satellite imagery. In some cases particular resources, including human and technology needed to execute the system may not be available due to either financial limitations or lack of political will. Such is usually the case for developing countries.

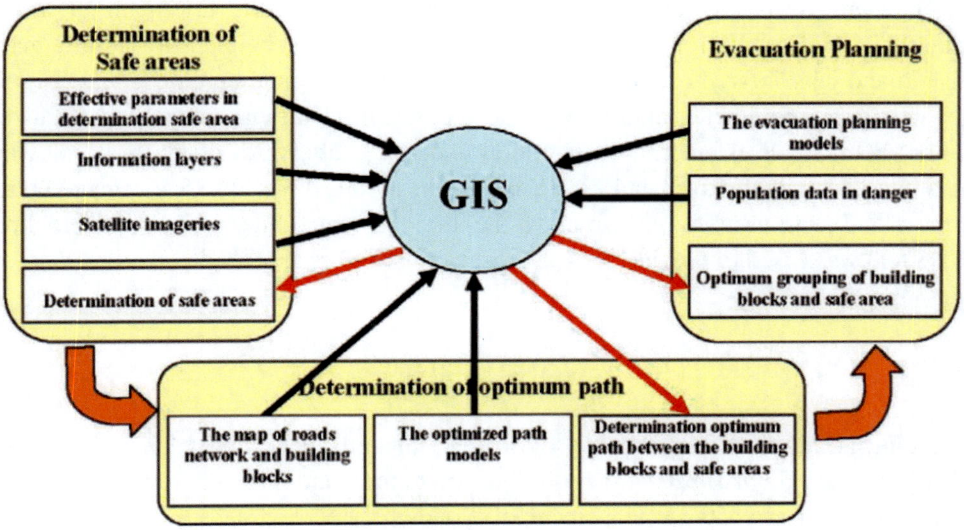

Figure 8 3. GIS evacuation plan (Naghdi et al., 2008).

One example in this respect is the island of Haiti where in January 2010 an earthquake of magnitude 7.0 struck resulting in massive destruction to property and killing over 100,000 persons. Relief efforts would have failed if not for the quick thinking of government and private organizations worldwide making available vast amounts of up-to-date mid- to high-resolution satellite imagery of the island. Using these images, relief organizations were able to better plan their routes using visual analysis as well as integrated GIS technologies and become more effective in their approach.

The technologies of remote sensing and also GIS play a critical role in the post-disaster and rehabilitation phase. In addition to offering rapid flood monitoring and assessment, satellite images of real flood events and information from the ground are useful for actual flood damage assessment, future flood mitigation planning, and validation of hydraulic analysis (Altan and Kemper 2010).

Such an integrated approach was used in the city of Muscat, Oman that was hit by a cyclone (Azaz 2007). GIS and satellite images were used to organize the damage information and the post-disaster census information, in addition to evaluating sites for reconstruction. In this study, two high-resolution satellite images (IKONOS) were utilized; one image before the cyclone and one after. The two images were geometrically corrected, and a change detection technique was then applied to identify and assess the damage (Azaz 2007).

Another study developed a methodology to combine the spectral, spatial and contextual information captured by IKONOS images to produce an accurate property damage map after flooding that affected villages in the southern part of the Netherlands in 1995 (van der Sande et al., 2003). The detailed land cover map, water depth, and known relations between water depth and property damage were all contributed to the creation of this map. The study confirmed that maps derived from high spatial resolution geospatial imagery are most accurate for flood simulations and actual property damage assessment (van der Sande et al., 2003).

The results of the two previous studies emphasize the importance of using remote sensing and GIS in the damage assessment phase as part of an effective disaster management plan.

However, there is usually an overwhelming cloud cover in the aftermath of a flood, particularly in the Caribbean islands. Thus, it is very difficult to obtain clear satellite image in the visible and near infared region. Having the capability to penetrate clouds and rain, radar images would thus be a better alternative in such cases. In fact, radar imagery was utilized to promptly obtain flood extent and assess flood damage in aftermath of the flood disaster in China along Changjiang River in 1998 (Yang et al., 2002). A RADARSAT image obtained after the flood was used for determining the flood extent while optical satellite images (LANDSAT) were used to provide the land cover status before the flood.

CONCLUSION

Flooding is the most common and most predominantly occurring geohazard in today's world. It accounts for a substantial proportion of human and economic distress with both direct and indirect impacts. Management of flood disaster can be successful only when detailed knowledge is obtained about the character and magnitude of the flood as well as the vulnerability of the people, residential properties, infrastructure, agricultural crops, and economic activities in the potentially vulnerable area.

This chapter has presented the use of Earth observation and Geographic Information System s as enabling technologies for assisting in a comprehensive disaster management system. Having reviewed several cases for their applications, these technologies have emerged as powerful tools for dealing with various aspects of flood prevention, preparedness, relief, and recovery efforts of flood disaster.

Earth observation and GIS techniques can significantly contribute to improving the efforts in modelling flood events, developing proper flood-mitigation strategies, and providing relevant agencies with important information for alleviating flood damage. They can supplement or complement other structural and non-structural measures in flood management systems. Extensive use of these technologies can prospectively create long term databases on flood proneness, risk assessment and relief management.

REFERENCES

Alkema, D. (2004). RS and GIS applications in flood forecasting. *Proceedings of the National Workshop on Flood Disaster Management: Space Inputs.* India. 57-59.

Al-Tahir, R., Baban, S. & Ramlal, B. (2006). Utilizing emerging geo-imaging technologies for the management of tropical coastal environments. *West Indian Journal of Engineering, 29* (1), 11-21.

Altan, O. & Kemper G. (2010). Spatial Information for disaster management using examples from Istanbul. In M. Konecny et al. (eds.), *Geographic Information and Cartography for Risk and Crisis Management,* Lecture Notes in Geoinformation and Cartography, Springer-Verlag Berlin.

Azaz, L. (2007). Using remote sensing and GIS for damage assessment after flooding, the case of Muscat, Oman after Gonu tropical cyclone: Urban planning perspective. *Proceedings of 15th International Conference on Urban Planning and Regional*

Development in the Information Society. Vienna, http://programm.corp.at/cdrom2010/papers2010/CORP2010_182.pdf

Badilla, R. A. (2008). *Flood Modelling in Pasig-Marikina River Basin*. Thesis. International Institute for Geo-Information Science and Earth Observation, Netherlands.

Ehlers, M. (2004). Remote sensing for GIS applications: New sensors and analysis methods. In Ehlers, M., Kaufmann, H., and Michel (eds.), *Remote Sensing for Environmental Monitoring, GIS Applications, and Geology III, Proc. of SPIE*, Vol. *5239*, 1-13.

Few, R., Ahern, M., Matthies, F. & Kovats, S. (2004). *Floods, Health and Climate Change: A Strategic Review*. Tyndall Centre Working Paper 63.

Hoyois, P. & Guha-Sapir, D. (2004). Disasters caused by flood: Preliminary data for a 30 year assessment of their occurrence and human impact. *Health and Flood Risk Workshop*. University of East Anglia. UK. http://www.em-dat.net/documents/working_paper/30yearsoffloods3.pdf

Johnson, R. (2000). *GIS Technology for Disasters and Emergency Management*. ESRI White Paper. ESRI.

Kafle T., Hazarika, M., Shrestha, K., Prathumchai, K. & Samarakoon, L. (2006). Integration of remote sensing and GIS with flood simulation model for flood hazard mapping in the Bagmati River, Nepal. *Proceedings of the 5th International Symposium on New Technologies for Urban Safety of Mega Cities in Asia,* Phuket, Thailand.

Khanna, R. K., Agrawal, C. K. & Kumar, P. (2006). Remote Sensing and GIS applications in flood management. *Proceedings of the India Disaster Management Congress,* New Delhi. pp. 11.

Li, R. (1998). Potential of high-resolution satellite imagery for national mapping products. *Photogrammetric Engineering and Remote Sensing*, *64*(12), 1165-1169.

Mansor, S., Abu Shariah, M., Billa, L., Setiawan, I. & Jabar, F. (2004). Spatial technology for natural risk management. *Disaster Prevention and Management*. Volume *13*, Number 5, 364-373.

Marfai, M.A. (2003). *GIS Modelling of River and Tidal Flood Hazards in a Waterfront City. Case study: Semarang City, Central Java, Indonesia.* MSc. Thesis, International Institute for Geo-Information Science and Earth Observation, Netherlands. Available online: http://www.itc.nl/Pub/Home/library/Academic_output (Last accessed on 10 May 2010)

Meyer, V., Haase, D. & Scheuer, S. (2009). A Multicriteria Flood Risk Assessment and Mapping Approach. In: Flood Risk Management: Research and Practice. Allsop, W., Samuels, P., Harrop, J. and Huntington, S. (Editors). Taylor and Francis Group, London, pp. 1687-1693.

Mori, M. (2007). GIS analysis of river flooding by high-resolution (2m) DEM. *Proceedings of the Annual Conference of the Remote Sensing &Photogrammetry Society*. Newcastle upon Tyne. UK

Murphy, A., Colleton, N., Downs, R., Goodchild, M., Hanson, S., Lawson, V., Macdonald, G., Magilligan, F., Moseley, W., Polsky, C., Seto, K. & Wright, D. (2010). *Understanding the Changing Planet: Strategic Directions for the Geographical Sciences*. National Research Council. The National Academies Press. Washington, D.C.

Naghdi, K., Mansourian, A., Valadanzoej, M. & Saadatseresht, M. (2008). Evacuation planning in earthquake disasters using RS & GIS. *The International Archives of the Photogrammetry, Remote Sensing and Spatial Information Sciences*. Vol. *XXXVII*. Part B4. pp. 1671-1676

Organization of American States (OAS) (1991). *Primer on Natural Hazard Management in Integrated Regional Development Planning*. Organization of American States, Washington, D.C. http://www.oas.org/osde/publications/Unit/oea66e/begin.htm.

Plate, E. J. (2007). Flood Risk Management for Setting Priorities in Decision Making. In: *Extreme Hydrological Events: New Concepts for Security*. Vasiliev, O.F., Van Gelder, P.H.A.J.M., Plate, E.J. and Bolgov, M.V. (eds). Vol. *78*, part 1, Springer Netherlands, New York, 21-44.

Pradhan, B. (2009). Flood susceptible mapping and risk area delineation using logistic regression, GIS and remote sensing. *Journal of Spatial Hydrology*. Vol. *9*, No.2, 1-18.

Ramlal, B. & Baban, S. (2008). Developing a GIS based integrated approach to flood management in Trinidad, West Indies. *Journal of Environmental Management*, *88* (4), 1131–1140.

Rango, A., Laliberte A. & Winters, C. (2008). Role of aerial photos in compiling a long-term remote sensing data set. *J. Appl. Remote Sens.*, Vol. *2*, Issue 1

Shanley, L. A., Bellovary, T., Lulloff, A. & Schwoegler-Boos, P. A. L. (2006). *GIS Data Sharing and Flood Hazards in Wisconsin.* Wisconsin Land Information Association, Wisconsin. Available online: www.wlia.org/resources /datasharingfloods.pdf. Last accessed 15 May 2010.

United States General Accounting Office (GAO) (2004). *Flood Map Modernization*. Report GAO-04-417. http://www.gao.gov/cgi-bin/getrpt?GAO-04-417.

USGS. (2007). Science Topics. Floods. The United States Geological Survey. Available online: http://www.usgs.gov/ (Last accessed on 25 April 2010).

van der Sande, C. J., de Jong, S. M. & de Roo A. P. J. (2003). A segmentation and classification approach of IKONOS-2 imagery for land cover mapping to assist flood risk and flood damage assessment. *International Journal of Applied Earth Observation and Geoinformation*, Volume *4*, Issue 3, Pages 217-229

Van Westen, C. J. (2002). Remote Sensing and GIS for natural disaster management. In A. Skidmore (ed.), *Environmental Modelling with GIS and Remote Sensing*. Taylor & Francis, London.

Wallingford, H. R. (2007). GIS -based Multicriteria Analysis as Decision Support in Flood Risk Management. Report number T10-07-07, FLOODsite Consortium, UK. Available on: www.floodsite.net (Last accessed on 25 April 2010)

Wilson, M. D. & Atkinson, P. M. (2003). A comparison of remotely sensed elevation data set s for flood inundation modelling. *Proceedings 7th International Conference on GeoComputation*, Southampton, UK (on CD-ROM).

World Water Assessment Programme (WWAP) (2006). *The 2nd UN World Water Development Report:* Water, a Shared Responsibility. UNESCO and Berghahn Books. Paris and New York.

Yang, C., Huang, H., Wei, Y., Zhu H. & Zhuo, J. (2002). Rapidly Assessing the Flood Disaster by Using Remote Sensing and GIS. *Proc. of IGARSS*, Canada, 2880-2882.

In: Flooding and Climate Change
Editors: D. D. Chadee, J. M. Sutherland and J. B. Agard
ISBN: 978-1-62948-317-7

Chapter 9

USING GIS IN DEVELOPING AN INTEGRATED APPROACH TO FLOOD MANAGEMENT IN TRINIDAD AND TOBAGO

Bheshem Ramlal*[*] *and Dexter Davis
Department of Geomatics Engineering and Land Management
The University of the West Indies, St. Augustine, Trinidad

ABSTRACT

In the last few decades in particular, Trinidad and Tobago has suffered a perennial flooding problem. Extensive physical development in most of the watersheds of the country has led to extensive flooding in low-lying areas. The problem is exacerbated when the tides are high during periods of heavy rainfall since the flood waters are prevented from flowing into the sea. Flooding has led to significant damage to livestock, agricultural produce, homes and businesses in many parts of the country, and, in some instances, loss of human life. Clearly there is a need for developing flood mitigation and management strategies to address flooding in the areas most affected. This chapter explores the use of GIS in developing flood mitigation and management strategies that may be employed in ensuring that the impacts of flooding are minimized.

1. INTRODUCTION

Flooding has become a major problem in many countries including India, the United States, Germany, the Netherlands, Guyana, and Trinidad and Tobago. Recent years have been marked by a significant increase in the number and severity of floods in many of the low lying areas of Trinidad and Tobago. These flooding events have led to significant damage to livestock, agricultural produce, homes and businesses and even to instances of loss of human life. In the last 20 years taking the country as a whole, floods caused the death of many persons, destroyed agricultural produce worth over US$30 million, and damaged over 2000

[*] Corresponding author: E-mail: bheshem.ramlal@sta.uwi.tt.

homes and businesses (ODPM, 2010). In the United States, the average annual loss to flooding is US$2.4 billion dollars (FEMA, 2006). Similarly in Germany, losses due to flooding for a four-year period exceeded 17.5 billion dollars (Conway 2006). While not as extensive, the annual average loss per household estimated in 1999 was US 1,000 dollars in some watersheds in Trinidad (Stere et al., 1999). Accounting for inflation and increased severity and frequency of flooding events, the present figure is likely to be much higher than the 1999 estimate. The overall impact of floods on Trinidad and Tobago is indeed significant.

In most watersheds in Trinidad and Tobago, the root of the flooding problem stems from the inappropriate use of lands including slash and burn agriculture, quarrying, illegal logging, forest fire s, and from the modification of the floodplain through both legal and illegal settlements. These have led to exposed soils that are more susceptible to erosion during periods of heavy rainfall and subsequent runoff. Consequently, heavy sedimentation occurs in the river channels causing these channels to be reduced in size. This ultimately leads to flooding. In addition, the lack of vegetative cover leads to much shorter lag times between rainfall and the water reaching the waterways causing the already reduced channels to overflow leading to massive floods (Dion, 2002).

Inappropriate land use is further exacerbated by several problems including the absence of relevant information on the level of vulnerability of different areas to damage, the lack of resources to address the problems of drainage maintenance, the lack of political will and inadequate resources for the relevant authorities to enforce the laws of the country (Smart, 2008) and the absence of appropriate legislation to prevent the extent of environmental degradation that is allowed to occur (Kumarsingh, 2008). The prospect of increased numbers of extreme climatic events due to global climate change is also likely to further intensify the situation, especially in the most vulnerable areas (Senior et al., 2002; Cubasch et al., 1999).

There is therefore an urgent need to introduce mitigation measures to ensure that, not only vulnerable areas are protected, but potential damage is avoided throughout Trinidad and Tobago. The first step in achieving this is to identify the nature and extent of vulnerability of all areas in the country. Second, identify the best available mitigation measures that may be used to address the problem. Finally, develop strategies for implementing and maintaining these measures.

2. Available Flood Management Practices

The US Federal Emergency Management Agency (FEMA) has developed a list of strategies and tools for flood management that recommend the following four major approaches (Melancon, 2003; Wright, 2008; FEMA, 2010):

2.1. Modify Susceptibility to Flood Damage and Disruption

The use of non-structural measures is employed in this strategy. For the most part, policy is employed to prevent and mitigate flood damage. Various policies that can be used to control and manage physical development are normally developed to target the affected community or region. In addition, strategies to retrofit or to construct new buildings to make them flood proof are also introduced. Flood prediction and early warning systems, emergency evacuation plans and disaster management and recovery plan s are also envisaged in this approach.

2.2. Modify Flooding

Structural measures are used to address the problems that already exist and to mitigate anticipated future problems. Several structures may be introduced in modifying floods. These include dams and reservoirs, dikes, levees, and floodwalls. Other approaches in modifying floods is to alter channels by deepening, widening, and straightening. In addition, spillways and high-flow diversions may be built. Land treatments such as adding vegetative cover, erosion control structures, terraces and crop management systems may also be used to modify flooding. A strategy applicable especially for new structures is the creation of onsite detention facilities. The site therefore contributes little to flooding.

2.3. Modify the Impact of Flooding on Individuals in the Community

In this approach individual landowners are targeted. Awareness through information and education are used to ensure that landowners are best able to prepare for floods. Strategies are also available to protect the lives of individuals and their properties. These include flood insurance, tax incentive s, emergency measures in the case of a flood, and post-flood recovery measures.

2.4. Protect and Restore the Resources and Functions of the Floodplain

When a floodplain is modified through increased development, it loses its functionalities with respect to the absorption of excess water and the filtering of pollutants (Melancon, 2003). The approach is to use education and policy to attempt the protection and restoration of the functions of the floodplain. See figure 9.1 for a summary of the strategies suggested by FEMA for use for flood management.

Based on these strategies it is possible to examine the use of geographic information systems in flood management in general, and it application to Trinidad and Tobago in particular.

3. GIS IN FLOOD MANAGEMENT

One of the best possible approaches for identifying appropriate flood mitigation strategies is to use spatial analysis tools available in Geographic Information System s (GIS) (Burrough, 1986; Bernhardsen, 1992). Spatial analysis tools have been developed to examine spatial and temporal patterns and to find associations between various geographical factors (Mitchell, 1999). Since flooding is a spatial phenomenon and is a consequence of a number of factors including slope, soil type, land use, vegetation cover and type, rainfall intensity, and rainfall frequency; the GIS allows the user to handle, manage and analyse the spatial data set s to determine which factors have what effect and to forecast the resulting consequences (Melesse and Shih, 2000).

A. Modify susceptibility to flood damage and disruption

1. Floodplain Regulations

(i)Zoning Regulations

(ii)Sub-division regulations

(iii)Building and Housing Codes

(iv)Sewer Regulations

(v) Other Regulatory Mechanisms

2. Development and Redevelopment Policies

(i) Services and Utilities

(ii)Land Rights, Acquisition, Open Space

(iii)Redevelopment and Urban Renewal

(iv)Evacuation/relocation

3. Disaster Preparedness, Response, Recovery

4. Flood proofing of structures

5. Flood prediction and early warning systems

B. Modify Flooding

Use of structural solutions such as:

1. Dams and Reservoirs

2. Dikes, Levees and Floodwalls

3. Channel Alterations

4. High-flow diversions and spillways

5. Land Treatment

6. Onsite Detention

7. Shoreline Protection Measures

C. Modify the impact of flooding on individual in the community

Use of the following strategies:

1. Information and Education

2. Flood Insurance

3. Tax adjustments

4. Flood Emergency Measures

5. Post Flood Recovery

D. Protect and restore the resources and functions of the floodplain

1. Education

2. Policy

Figure 9.1. Flood Management Strategies (adapted from Wright 2008 and FEMA 2010).

Table 9.1. A Summary of Flood Management Strategies and the appropriateness of GIS

Flood Management Strategy	Application of GIS
Zoning Regulations	GIS may be used to undertake analysis to classify areas into different flood zones. Maps that show how each area is likely to be affected under different physical and climatic conditions may be generated. GIS may also be used to monitor changes and to generate scenarios of future events
Sub-division regulations	GIS may be used to determine the limitations associated with proposed sub-divisions based on the flood zones identified above. Regulations limiting the development of plots that are located in these zones may then be introduced
Building and Housing Codes, Sewer Regulations	GIS may be used in predicting the likely hazards that buildings may encounter in different areas and can be used to assist decision makers to develop building/housing codes appropriate to each of these areas. Similarly, the development of regulations for sewer systems may be assisted using GIS.
Services and Utilities Policies	The nature and extent of the services and utilities needed may be identified using GIS. The basis for the requirements f or developing policies may be established.
Land Rights, Acquisition, Open Space	GIS may be used to record all rights in land including information about land use, vulnerability to hazards including flooding. Policies governing all lands may be guided by the results of spatial analysis executed using the tools available in GIS.
Redevelopment and Urban Renewal	Spatial databases can be developed and used to undertake analysis to identify areas for redevelopment and urban renewal. In addition, policies that govern redevelopment and renewal may be developed based on the spatial data available in GIS.
Evacuation/ relocation	GIS may be used to assess the flooding risk associated with all areas under consideration. Based on this assessment, all stakeholders that are likely to be affected can be identified and plans for evacuating them can then be developed. Target areas for relocating affected stakeholders can also be identified. GIS allows for holistic and comprehensive planning to be conducted.
Disaster Preparedness, Response, Recovery	GIS may be used in every aspect of disaster preparedness, response and recovery to generate plans for ensuring that all aspects of preparation i s considered and accommodated, response is coordinated and recovery is planned and executed in the most effective and efficient manner possible given the human and material resources available to the stakeholders
Flood proofing of structures	GIS may be used to analyze all areas and structures that may be vulnerable to flooding and the extent to which these areas are likely to be flooded. The type of flood proofing needed can then be identified.
Flood prediction and early warning systems	GIS provides powerful tools for the generation of "what if" scenarios and therefore is most appropriate for flood prediction. In addition, GIS may be coupled with flood monitoring systems and early warning systems to provide an optimum solution to protect life and property in areas that may be at risk from flooding.
Modify Flooding through Structural Measures	GIS is ideal for the identification of the need for, and the location of, various structural measures that may be used for flood mitigation. The size and capacity of the identified structure may also be determined using GIS
Information and Education, Flood Insurance, Tax adjustments, Flood Emergency Measures, Post Flood Recovery	Based on the spatial data gathered, it is possible to provide information for dissemination to stakeholders. This information may also be used for educating stakeholders on activities that may reduce or contribute to flooding. GIS may be used to record the flood reduction and prevention measures taken by individuals and communities. These may be used to determine flood insurance rates, compensation, and tax incentives. In addition, the GIS may be used for developing emergency m easures and post-flood recovery strategies.
Protect and restore the resources and functions of the floodplain through Education and Policy	GIS may be used to generate information for the development of education a nd policy formulation purposes. The availability of information and communication t echnologies makes it easier to provide the necessary support so that stakeholders are part of the process to ensure the best possible strategies are adopted.

In addition, GIS provides the ability to carry out temporal analysis, which is essential for flood prediction. By storing data on previous floods, soil types, and river channel size, it is possible to create a model of peak flow, discharge and runoff. In terms of the impacts of land use and land cover on flooding, GIS can be used to detect change as well as identify trends, both visually and statistically, between land use changes and flooded areas (Mamat and Mansor, 1999). It may be noted that there are several types of hydrologic models that may be used for predicting erosion, sediment transport and flooding (Summer et al., 1998).

Geographic information systems have been used in various aspects of flood analysis. See Zhiyu (2009) for a discussion on an integrated urban flood-management strategy for China and the use of GIS in flood mitigation and management. Others have put forward geo-spatial solutions for disaster management, including flooding, for the city of Mumbai, India (Purohit and Mantri 2009). Many researchers have reported the use of GIS for various specific aspects of flooding including the compilation of spatial data set s, quantitative analysis of flooding, using GIS to evaluate structural measures to be used in flood mitigation, and the social and ecomomic impacts of flooding-mitigation measures. See for example: Zheng et al. (2009), Carlin, (2009); Qiu et al. (1997); and Msuya, (2009).

In Trinidad and Tobago, GIS has been used for flood analysis in the Caparo River Basin (Stere et al., 1998, 1999; Ramlal and Baban, 2008), Other recent studies using GIS for flood and erosion analyses are reported in Ramlal (2008), Baban (2008) and Cooper et al. (2008).

Table 9.1 provides a summary of the applicability of GIS in various aspects of flood mitigation and management. This table uses the strategies presented in figure 9.1 and provides examples of GIS applications for each based on information gathered from a review of the literature available on flooding. It may be noted that the list of examples provided is not exhaustive but reflect the diverse spatial analysis capabilities available in GIS.

4. A GIS Approach to Flood Management in Trinidad and Tobago

Trinidad and Tobago is a twin island state located in the southernmost part of the Caribbean chain of islands. See Figure 9.2. Trinidad is approximately 5,000 square kilometers while Tobago is approximately 300 square kilometers in area. The two islands have a combined population of approximately 1.4 million people.

In providing solutions to address flooding in Trinidad and Tobago, it is necessary to understand the spatial characteristics of the islands. To avoid repetition, the following discussion focuses on the island of Trinidad only.

Trinidad consists of two plains separated by three ranges of hills (see Figure 9.3). The plains are the Oropuche plain and the Caroni plain. The hill ranges are the Northern Range that runs all along the northern part of the island from east to west. The highest hills are located in this range. The Central Range is on average much lower than the Northern Range and runs north-east to south-west. The Southern Range is lower than the Central Range and is significantly smaller is geographical extent.

Figure 9.2. Trinidad and Tobago in the Caribbean Context.

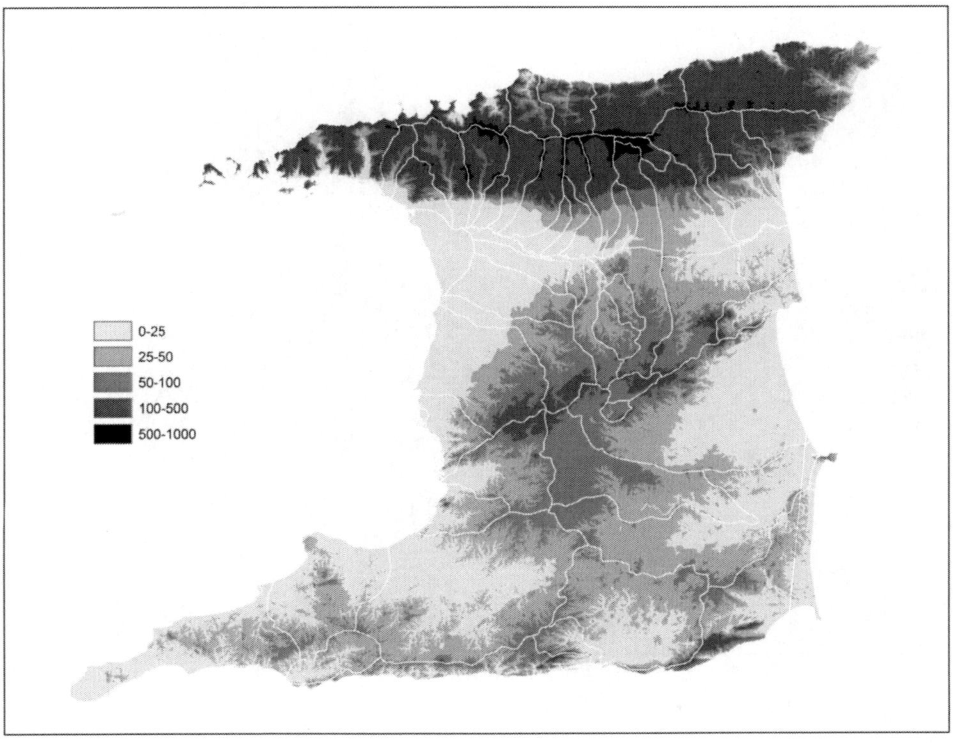

Figure 9.3. Elevation map (meters) of Trinidad with watershed boundaries.

In addition, there are two major swamps located on the island of Trinidad : the Caroni Swamp in the north-west of the island and Nariva Swamp in the north-east of the island. Most of the waterways in Trinidad are relatively small. The largest river, the Caroni, is approximately 20 meters wide and runs along the southern foothills of the Northern range in an east-west direction. Most of the larger rivers flow in an east west direction either through the plains to the coast or to the swamps.

The largest amounts of rainfall occur in the eastern part of the Northern Range and the least occurs in the western parts of the island. Figure 9.4 provides a distribution map of the average annual rainfall for Trinidad. It may be noted from an examination of Figures 9.3 and 9.4 that much of the rainfall occurs in the north-eastern part of the island and most of the water yielded drains into the Caroni river basin located south of the northern range and running parallel to it running from east to west.

An examination of the land use map of Trinidad show that most of the populated areas is located in the valleys of the Northern and Central Ranges and in the low lying areas of the Caroni and Oropuche plains on the Western parts of the island (See figure 9.5). Spatial data set s including soil type, rivers and water bodies, roads and bridges, and environmentally sensitive areas may be used together with the above data to undertake various types of analysis needed to develop a flood management plan for Trinidad and Tobago. The next sections review the major components of such as plan.

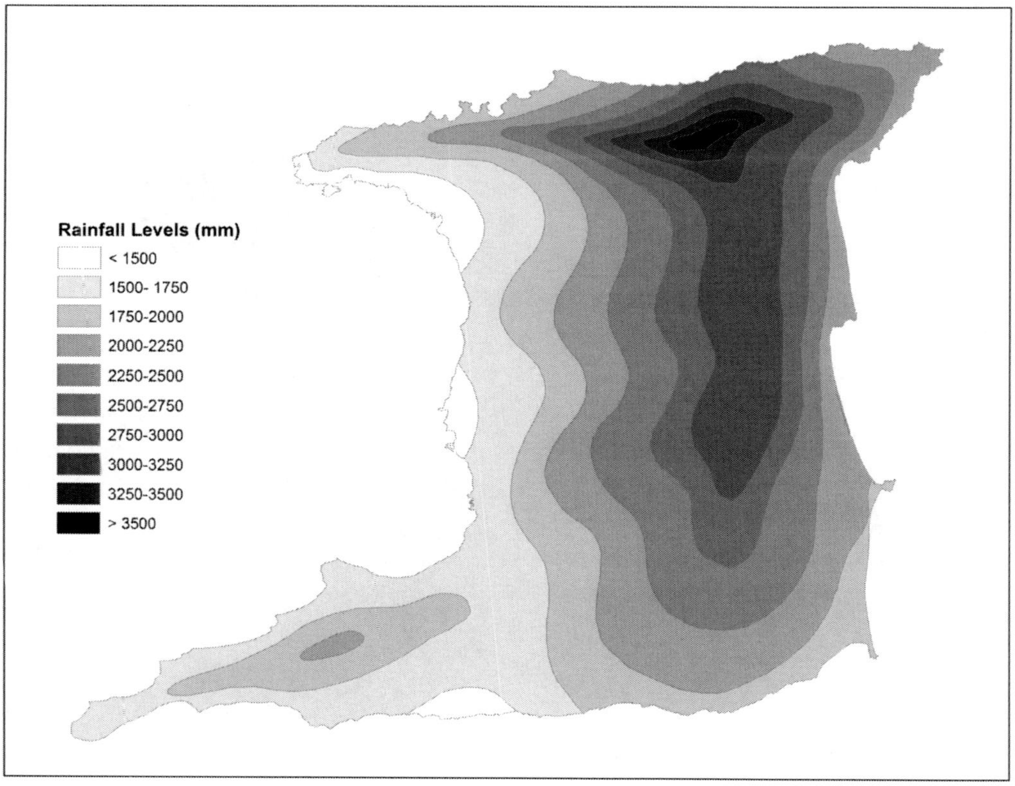

Figure 9.4. Average Annual Rainfall for Trinidad.

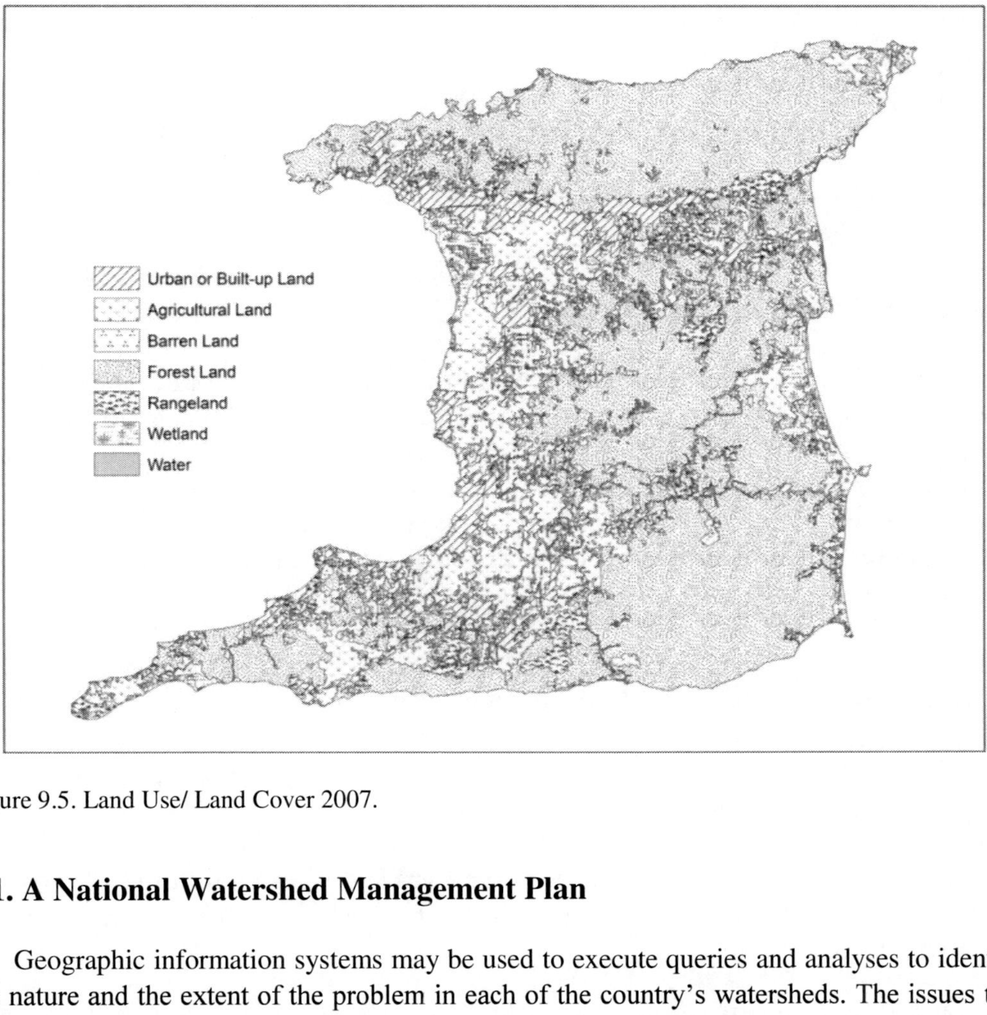

Figure 9.5. Land Use/ Land Cover 2007.

4.1. A National Watershed Management Plan

Geographic information systems may be used to execute queries and analyses to identify the nature and the extent of the problem in each of the country's watersheds. The issues that can be identified may include: the need for reforestation of steep slopes; the introduction of conservation measures for areas used for agricultural purposes; the need to control further development in the lower and middle reaches especially in flood-prone areas.

GIS can be used to identify areas for conservation based on spatial characteristics such as slope and soil type s. Conservation measures may include the use of earth bounds, contour drains, contour planting, vegetable barriers, diversion channels, graded drains, and strip cropping. This is normally appropriate for the upper reaches of a river basin. In the middle and lower parts of a river basin the major issue will most likely be that of expansion of development into the floodplains. Further development needs to be monitored and tightly controlled by the relevant agencies. Institutional arrangements need to be developed to ensure that agencies work in concert in managing watersheds.

4.2. A National Plan for Flood Control Works

The flood control works that may be proposed can include the upgrading of channels to increase conveyance capacities of channels, the construction of a retention pond in the watersheds to hold back the high discharges that enter the flatter valley floor to release at a rate that is suitable for the upgraded downstream channel capacity. In addition, bypass channels around the populated areas may have to be developed to divert the greater part of the flood-flows away from these areas. Trash racks can be placed upstream of populated areas so that trash is collected before the flow enters the urban area. Finally, lands can be allocated to store and convey floodwaters on rare occasions when floods are exceedingly high.

4.3. National Land Acquisition Plan

Several tracts of land may be required to install the diverse infrastructure needed to mitigate flooding in all river basins that are affected. Corridors along the general course of affected rivers may be needed for engineering works. In addition, areas may be needed to construct detention basins. A land acquisition strategy and relocation plan for any existing occupants can be developed to secure the lands needed. This strategy may be developed using GIS to manage cadastral interests along watercourse corridors as well as finding the ideal locations for detention ponds.

4.4. National Environmental Management Plan

Mitigation measures that are proposed must seek to minimize the negative impacts and maximize the beneficial impacts. Mitigation measures such as increase in vegetative cover through reforestation, more educated and effective use of pesticides to mitigate negative impacts; and the dredging and construction work in the dry season to avoid siltation and sediment transport due to runoff may need to be developed to ensure that environmental impact s are minimized.

4.5. A National Operation and Maintenance Plan

The above plans will require that major investments are made. It is therefore necessary to ensure that there is continued effectiveness in the infrastructure being introduced all to the affected areas. A plan to operate and maintain all facilities must be developed. For example, if a dam is built it is important that a dam safety unit is established with personnel who have appropriate training and resources to conduct all aspects of dam safety and maintenance monitoring. The plan should provide adequate resources and develop strategies to ensure that detention basins, river alignments, embankments, river crossings, diversion structures and flap gates are all well maintained.

CONCLUSION

Flooding in Trinidad and Tobago is caused by many factors including exposed soils, quarrying activities, slash and burn agriculture and silted drains. GIS may be used in determining the extent of the problem including the identification of areas that need conservation , and the extent of soil loss and flooding at different sections of watersheds. The analysis provided by GIS makes it possible to expedite the development of mitigation and management strategies that are most likely to address the anticipated changes in river basins.

REFERENCES

Baban, S. M. J. (2008). *Enduring Geohazards in the Caribbean : Moving from the Reactive to the Proactive*, The University of the West Indies Press.

Burrough, P. A. (1996). *Principles of Geographic Information System s for Land Resources Assessment*, Clarendon Press, Oxford.

Bernhardsen, T. (1992). *Geographic Information System s*, Viakit, Norway.

Carlin, N. (2009). Spatial Analysis Using Geographic Information System s (GIS) to Evaluate Areas Susceptible to Repeat Flash Flooding in La Crosse County, Wisconsin. Volume *11*, Papers in Resource Analysis. 12 pp. Saint Mary's University of Minnesota Central Services Press. Winona, MN. Retrieved (date) from http://www.gis.smumn.edu

Conway, J. (2006). *Flood in Europe : A Peril Not to be Ignored.* www.usau.com/facworld.nsf/ Doc/euflood/$file/floodeu2.pdf, Accessed August 2006.

Cooper, V., Ramlal, B., Phelps H., Farabi, H. & Gay, D. (2008). Nariva Restoration Project: Final Report, Environmental Management Authority, Unpublished Report.

Christensen, J. H., Krishna Kumar, K., Aldrian, E., An, S.-I., Cavalcanti, I. F. A., de Castro, M., Dong, W., Goswami, P., Hall, A., Kanyanga, J. K., Kitoh, A., Kossin, J., Lau, N.-C., Renwick, J., Stephenson, D. B., Xie, S.-P. & Zhou, T. (2013). Climate Phenomena and their Relevance for Future Regional Climate Change. In: *Climate Change 2013: The Physical Science Basis.* Contribution of Working Group I to the Fifth Assessment Report of the Intergovernmental Panel on Climate Change [Stocker, T.F., D. Qin, G.-K. Plattner, M. Tignor, S.K. Allen, J. Boschung, A. Nauels, Y. Xia, V. Bex and P.M. Midgley (eds.)]. Cambridge University Press, Cambridge, United Kingdom and New York, NY, USA.

Dion, T. R. (2002). *Land Development for Civil Engineers*, 2nd Edition, Wiley, New York.

FEMA (2006). Floodsmart.gov *Flood Statistics.* http://www.floodsmart.gov/floodsmart/ pages/fastfacts.jsp, Accessed August 2006.

FEMA (2010). Floodplain Management Publications, http://www.fema.gov/plan/prevent/ floodplain/publications.shtm, Accessed June 2010.

Kumarsingh, K. (2008). Former Environmental Officer, Environmental Management Authority, Government of Trinidad and Tobago. Personal Communications.

Mamat, R. & Mansor, S. B. (1999). Remote Sensing and GIS for Flood Prediction. Proceeding of Asian Association of Remote Sensing, 5 pages, http://www. gisdevelopment.net/acrs/1999.

Melancon, J. (2003). Flooding, Tier 2 Worksheet, Community Environmental Management, Millbrook, New York. Draft Report. http://www.ecswcd.org/cem/t2b_flood.pdf

Melesse, A. M & Shih, S. F. (2000). Geomorphic GIS Database for Runoff Coefficient Determination. *Proceedings of the Second International Conference on Geospatial Information in Agriculture and Forestry*, 10-12th January 2000, Lake Buena Vista, Fl. Vol. *1*, 505-512.

Mitchell, A. (1999). The ESRI Guide to GIS Analysis, Vol. *1*. ESRI Press.

Msuya, O. (2009). Using GIS to Analyze the Economic and Social Benefits of a Flood Control Levee in Mankato, Minnesota. *Papers in Resource Analysis* 11, 12 pp. Saint Mary's University of Minnesota Winona, MN. Retrieved (date) http://www.gis.smumn.edu.

ODPM (2010). Office of Disaster Preparedness and Management, Ministry of National Security, Government of Trinidad and Tobago, Personal Communication.

Purohit, S. & Mantri, S. (2009). Formulation of Geospatial solution for disaster management. International Conference on Geo-spatial Solutions for Emergency Management and the 50th Anniversary of the Chinese Academy of Surveying and Mapping, Beijing, China, 14-16 September 2009, *International Archives of Photogrammetry, Remote Sensing and Spatial Information Sciences* 38, 7/C4, Editors: Zhang Jixian, Li Zhilin, Cheng Pengfei, Yan Qin, http://www.isprs.org/proceedings/XXXVIII/7C4/67_GSEM2009.pdf

Qiu, H., Luna, M. & Zagurski, T. (1997). Compilation of GIS Data Sets for Flood Control Alternatives in California, Final Report, Department of Geography and Urban Analysis, California State University, Los Angeles

Ramlal, B. (2008). "Using GIS for Flood Management and Mitigation in Trinidad and Tobago " In Serwan M.J. Baban, editor *Enduring Geohazards in the Caribbean : Moving from the Reactive to the Proactive*, Chapter 6. The University of the West Indies Press.

Ramlal, B. & Baban, S. M. J. (2008). Developing a GIS based integrated approach to flood management in Trinidad, West Indies, *Journal of Environmental Management*, *88*, 1131-1140.

Senior, C. A., Jones, R. G., Lowe, J. A., Durman, C. F. & Hudson, D. (2002). Predictions of extreme precipitation and sea-level rise under climate change, *Philosophical Transactions of the Royal Society Londdon A*, (2002) *360*, 1301-1311.

Smart, C. (2008). Former Director, Town and Country Planning Division, Ministry of Planning and Development, Government of Trinidad and Tobago, *Personal Communication.*

Stere, C., Hesterman, E. & Passchier, A. (1998). Caparo River Basin Flood Mitigation and Water Resources Management Project: Problem Identification and Analysis Report. Unpublished Report, Government of Trinidad and Tobago.

Stere, C. Hesterman, E. & Visser, S. J. (1999). Caparo River Basin Flood Mitigation and Water Resources Management Project. *Unpublished Report*, 4, 69pp. Government of Trinidad and Tobago.

Summer, W., Kaghofer, E. & Zhang. W. (1998). Modelling Soil Erosion, Sediment Transport and Closely Related Hydrological Processes, IAHS Publication no. 249. IAHS Press, Oxfordshire.

Wright, J. M (2008), Floodplain Management, Principles and Current Practices, University of Tennessee. http://training.fema.gov/EMIWeb/edu/fmpcp.asp

Zheng, N., Takara, K., Yamashiki, Y. & Tackikawa, Y. (2009). Large Scale Quantitative Vulnerability Analysis for Regional Flood Hazard, *Annual of Dis. Prev. Res. Inst. Kyoto Univ*. No.52 B 2009.

Zhiyu, L. (2009). Strategies and Countermeasures for Integrated Urban Flood Management in China, Expert *Group Meeting on Innovative Strategies Towards Flood Resilient Cities in Asia -Pacific* UNESCAP, 21-23 July 2009, Bangkok, Thailand, http://www.unescap.org/idd/events/2009_EGM-DRR/index.asp

In: Flooding and Climate Change
Editors: D. D. Chadee, J. M. Sutherland and J. B. Agard
ISBN: 978-1-62948-317-7

Chapter 10

FLOODING IN JAMAICA: CAUSES AND CONTROLS

Michael A. Taylor[1,*], Arpita Mandal[2], Christopher Burgess[3] and Tannecia Stephenson[1]

[1]Climate Studies Group, Mona; Department of Physics, The University of the West Indies, Mona, Kingston, Jamaica
[2]Department of Geography and Geology, The University of the West Indies, Mona, Kingston, Jamaica
[3]CEAC Solutions Limited, Jamaica

ABSTRACT

Flooding from high intensity rainfall events, some associated with tropical storms and hurricanes, is a major hydrometerological hazard affecting Jamaica. Repeated flooding has been costly both in terms of lives and livelihoods, and flood frequencies have increased towards the end of the 20th century relative to previous decades. Flooding in Jamaica is a combined effect of rainfall, topography and lithology. Most floods occur in the months of May and September-October when rainfall peaks, with hurricanes and tropical storms accounting for approximately 43% of all the devastating flood events. The watersheds of eastern Jamaica, which experience some of the highest rainfall totals for the island, are particularly susceptible. The strong association with rainfall suggests some modulation by large-scale climatic events such as the El Niño Southern Oscillation (ENSO) and the longer term impact of climate change.

Flooding is of five types. Riverine flooding is predominant in the eastern section of the island where steep slopes and the presence of impermeable lithology (volcanic and volcanoclastics) accommodate higher surface runoff than infiltration. Groundwater flooding is dominant in the central karst limestone terrains, and in the depression basins in the parish of Manchester. Urban flooding dominates in Kingston, primarily in the downstream sections of the Hope river watershed, Liguanea plains and St Catherine, while coastal flooding has been experienced around the entire coastline of Jamaica. Extreme storm surge levels (2.5 - 4.5 metres) are observed during the passage of tropical storms and hurricanes.

* Corresponding author: Michael A. Taylor. E-mail: Michael.taylor@uwi.mona.edu.

Jamaica has a history of using hydrological modelling, particularly, with respect to the planning and development of large and newer infrastructure in designated flood-prone areas. More recently there has been an additional emphasis on modelling to also determine the vulnerabilities of communities in the major watersheds to flooding under present and future climatic conditions. An example of recent hydrological modeling in eastern Jamaica is presented. There is recognition of the need for flood hazard management. Employed methodologies include the use of early warning systems and flood plain maps, and the allocation of specific management responsibilities to various state entities.

1. Introduction

Jamaica is the third largest island in the Caribbean Sea. It is centred on latitude 18°15' N and longitude 77°20' W, and is approximately 145 kilometres south of Cuba and 145 km west of Hispaniola. It has a total landmass of 10,991 square kilometres and approximately 1,022 kilometres of coastline. It is elongated along a west-northwest to east-southeast alignment, and is approximately 230 kilometres long and 80 kilometres wide at its broadest point. Jamaica has several rugged mountain ranges which dominate its inland and a narrow coastal plain on which most of its major cities and towns and critical infrastructure is located.

Jamaica 's highest point, the Blue Mountain Peak, is a part of the Blue Mountain range and reaches over 2,256 metres (7,402 feet). The island has a Cretaceous volcanic-sedimentary basement, which is overlain by the Yellow limestone and White Limestone groups. The Central Inlier has eroded through these to expose the Cretaceous rocks.

About sixty percent of the island's bedrock is white limestone ; twenty five percent is volcanic and cretaceous, ten percent alluvial and five percent yellow limestone. More than 120 rivers flow from the mountains to the coast.

There are fourteen parishes in Jamaica, with Kingston being the capital of the country. The climate of Jamaica is mainly tropical with large scale weather circulation features and the island's orography contributing to a dry winter-wet summer rainfall pattern.

2. Flooding As an Economic Hazard

Records of severe flooding events and the toll on Jamaican society exist from 1678. In addition to newspaper archives, several national agencies have historically undertaken the responsibility of assessing and recording flood events, including: the Office of Disaster Preparedness and Emergency Management (ODPEM) Flood Catalogue[1]; the Meteorological Services (MS) record of Tropical Cyclones; the Water Resources Authority (WRA); and recently the Planning Institute of Jamaica (PIOJ).

Figure 10.1a and b show the distribution per year and per 10-year period from 1904[2] of all reported flooding events (irrespective of severity) from ODPEM and newspaper archives.

[1] ODPEM (2000) Disaster Catalogue: Flood catalogue.

[2] Although data has been archived from 1692 to 2012 from both ODPEM and newspaper archives, exact dates of flooding events along with the areas of flooding are fully achievable from 1904 to 2012.

Flood frequencies have increased at the end of the 20th century and into the 21st century with the periods 1956 to 1965 and 1996 to 2005 having maximum occurrences of events.

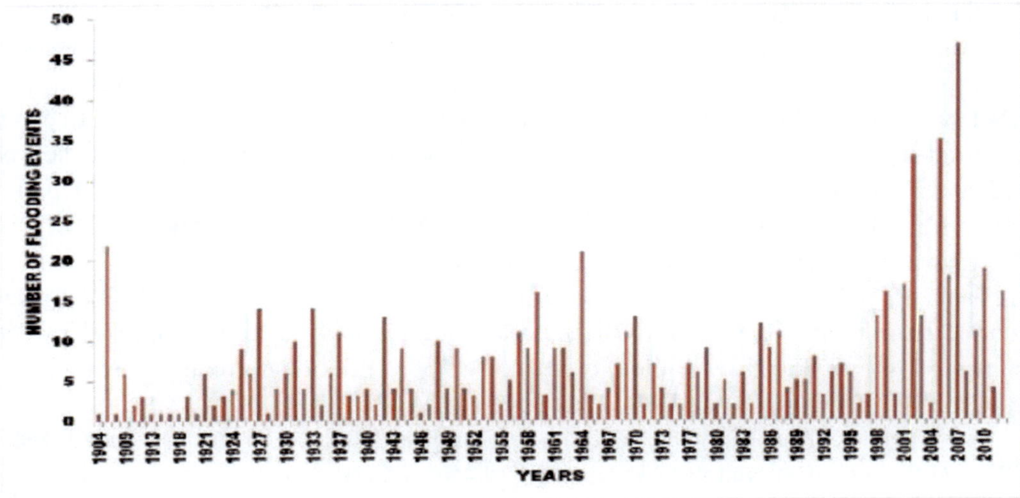

a

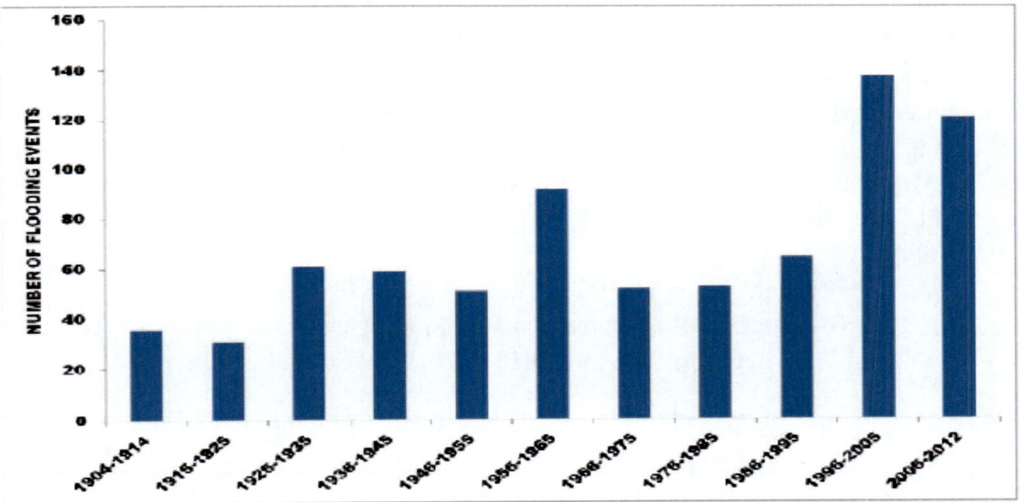

b

Data from Mona Geoinformatix Ltd, MGI.

Figure 10.1. (a) The number of flooding events between 1904 to 2012. (b) Flood events per 10 year period.

If the events are restricted to only those reported by the agencies above during the years 1900 to 2010, there is evidence of a 20 to 30 year cycle, a mean of 18 severe events per decade, and a strong increase in severe flooding since 1990 (103 events). (Severity in this sense is determined solely by the fact that the event is reported on by a national recording entity). The long cycle hints at long term modulation by a multi-decadal climatic feature, while the recent increase in both magnitude and frequency may hint at a long term climatic signal.

The latter increase could, however, also be attributable in part to increased documentation commensurate with greater reporting ability and enhanced institutional capacity.

The economic impact of flooding and extreme precipitation events is reported on by the Planning Institute of Jamaica (PIOJ) in their Damage and Loss Assessment Reports[3].

Analysis of these reports suggests significant recurring and negative impact on Jamaica 's economy due to flooding. The Planning Institute of Jamaica in their Vision 2030 Development Plan (PIOJ 2009) noted that during the first decade of the twenty-first century, hydro-meteorological hazards (flooding and other related impacts) from six storms between 2002 and 2007 resulted in damages and losses amounting to approximately $70.72 billion (in 2007 prices) or approximately 3.2 per cent of current Gross Domestic Product (GDP) (i.e., relative to a 2010 GDP value of JMD$1.018Trillion[4]).

Table 10.2 provides the distribution of the reported losses due to the 55 flood events reported on in the 114 years of PIOJ record keeping in 2010 Jamaican dollar value. Total economic loss is JMD$287.0 Billion, or about JMD$5.2 Billion on average. Median loss is between JMD$100 Million and JMD$10.0 Billion.

Events such as Hurricane Ivan (2004) and Hurricane Charlie (1951) account for JMD$71 Billion (5.8% of GDP) and JMD$53 Billion (4.4% of GDP) in losses respectively. In the context of the 2010 value, a typical significant flood event represents about 0.5% of GDP, so that an annual occurrence of two flooding events results in losses equivalent to about 1% of the GDP or JMD$10.4 Billion (2010 dollars).

Flood losses are largely accounted for by damage to infrastructure and loss of housing stock. For example, in St. Thomas, 10,000 people were left homeless after Hurricane Charlie (1951)[5] and 551,340 persons were affected by the floods associated with the May 1991 rains[6].

The May 1991 rains saw the island receiving a 24-hour rainfall of 20-38 cm which led to significant losses (Ahmad and Carby, 1993) in the agricultural sector and through damage to infrastructure.

Table 10.2. Distribution of Economic cost of flooding events in Jamaica reported between 1896 and 2010 in 2010 JMD$ dollar

Upper limit of Cost of Damages and Losses (2010 JMD)	Occurrences	Percentage occurrences
$ -	0	0%
$ 1,000,000.00	1	2%
$ 10,000,000.00	4	7%
$ 100,000,000.00	14	25%
$ 1,000,000,000.00	16	29%
$ 10,000,000,000.00	12	22%
$ 100,000,000,000.00	8	15%
$ 1,000,000,000,000.00	0	0%

[3] Planning Institute of Jamaica (2012) Damage and Loss Assessment Reports, http://pioj.gov.jm/ResearchandData/tabid/82/Default.aspx.

[4] Planning Institute of Jamaica (2010), Selected Indicators, website: http://www.pioj.gov.jm/Portals/0/Economic_Sector/SELECTED%20INDICATORS%202005%20to%202010.pdf.

[5] Jamaica Gleaner published August 11, 2004. http://jamaica-gleaner.com/gleaner/20040811/lead /lead2.html.

[6] The OFDA/CRED International Disaster Database. October 30, 2004.

3. INFLUENCING FACTORS

3.1. Flooding and Rainfall

Flooding in Jamaica is a combination of rainfall, topography and the lithology of the island. Among these, rainfall is the dominant factor. To a large extent the occurrence of flooding is strongly tied to the rainfall climatology and the features that give rise to rain at a particular time of the year. The rainfall climatology of Jamaica is a bimodal pattern with an early rainfall season centred in May and a late rainfall season centred in October (Figure 10.2a). A brief drier period in July separates the early and late wet seasons, and is referred to as the midsummer drought (MSD). Approximately 70% of total annual rainfall received by Jamaica falls between May and November, and approximately 40% falls between August and November (the late wet season). Rain during the wet season originates from convective systems embedded in passing easterly waves, tropical depressions, storms and hurricanes. The peak in the late season coincides with the peak in Atlantic hurricane activity. The beginning and ending months of the rainy season are also subject to the passage of cold fronts from the north. The dry season (December through March) is not completely dry as it too is influenced by the intrusion of cold fronts from the north.

Nkemdirim and Jones (1978), in an early analysis, showed that approximately half of the annual floods in the island occur in October, with the late wet season (September, October and November) accounting for two-thirds of all such events.

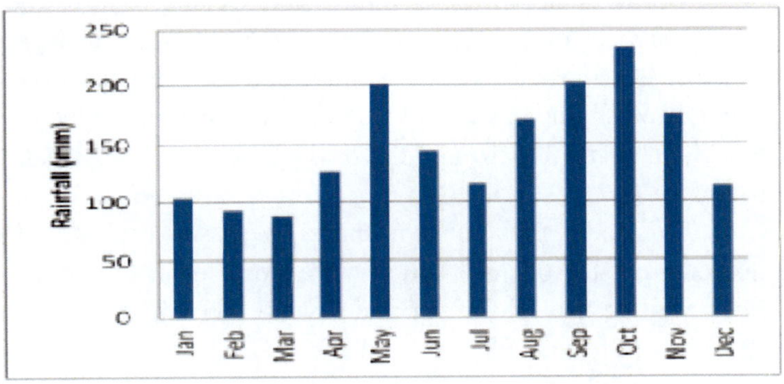

a

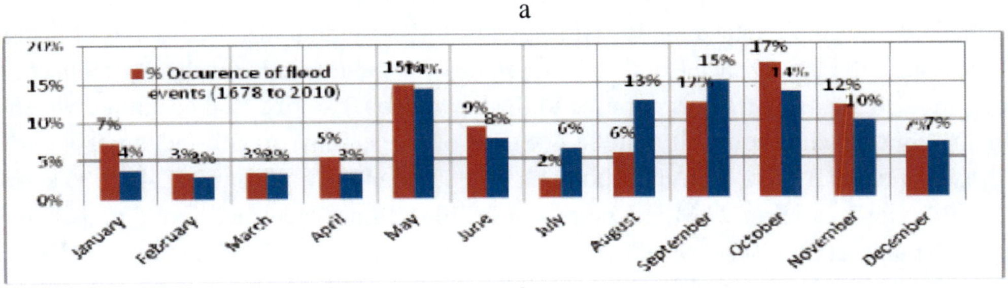

b

Figure 10.2. (a) Rainfall climatology of Jamaica. Base period: 1971-2000. Data source: Meteorological Service of Jamaica. (b) Occurrence of flood events (1678 to 2010, ODPEM, Meteorological Service) (red) and maximum one day rainfall (1992 to 2008, Meteorological Service) (blue) by month.

As with rainfall, there is a secondary peak in flood events in May, which accounts for about 12 per cent of flooding events in the analysis of Nkemdirim and Jones (1978), while the remaining months (i.e., outside of the wet season) have significantly less events. In the data analysed, July (the time of the MSD) was one of the least preferred months for floods. Analysis of recent rainfall data 1992 to 2008[7] and flood records from 1678 to 2010[8] reveals similar results (Figure 10. 2b). Peak occurrence of floods is seen in the months of October (~17% of recorded events) and in September (15% of recorded events). The early rainy season in May accounts for only 15% of the recorded events.

Floods also appear to be correlated with extreme precipitation. Frequency analysis of the maximum one day rainfall events by month shows the dominance of the late season, followed by that of the early season (Figure 10.2b).

Flooding events in Jamaica are associated with the range of rain making features. Frequency analysis of the reported causes of severe flood events indicates hurricanes as responsible for approximately 43% of all the devastating flood events, with troughs, tropical waves, tropical depressions, tropical storms and fronts accounting for the remaining and majority cases (56%) (See Figure 10.3). The October peak in flooding corresponds with the late hurricane and tropical storm seasons, and the increase in severe flooding events over the past decade can be directly correlated with the number of tropical storms /hurricanes which have hit or passed Jamaica since 1900. The number of storms and hurricanes influencing Jamaica has increased in frequency from 2000 to present (Figure 10.4). In any given year flooding may, however, be the result of any one or more of the rain producing phenomena. For example, from Figure 10.1 2002, 2007 and 2010 were years of multiple flood events.

In 2002 flooding occurred in May, September and October due to the presence of a surface to low level trough, a cold front and the passage of tropical depression fourteen. Flooding in 2005, 2007, 2010 was also attributable to intense rains from cold fronts/surface troughs (in the month of May), the passage of tropical storms Wilma (October 17, 2005) and Nicole (September 28-October 3, 2010) and the passage of hurricanes Dean (August 18-20, 2007) and Felix (Sept. 3, 2007)[9].

Though most of the flood events occur in the rainy season it is important to note that flooding events do occur outside of (or at the very start or end of) the peak flooding periods (most times in early May or late November and also in the dry season). These events are usually associated with cold fronts. Nkedrim and Jones (1978) note that "systems associated with cold fronts from mainland southeastern United States tend to stagnate over the Caribbean during the winter months, [giving rise] to heavy rainfall of the persistent variety. Consequently, the ability of the soil to infiltrate the incipient rain is drastically reduced, and floods are generated." This is as opposed to the convectional storms of the summer which are "usually isolated with the result that the dry soils of the summer months adequately cushion their flood potential." For example, flooding in Port Maria on November 24, 2006 resulted from a 24 hour rainfall of 240 mm due to a cold front. Water levels rose as high as 2.5 m in the town. Laing et al. (2004) also document the impact of a flash flood due to a cold front in January 3-4, 1998 (during the dry season) which caused six deaths and over nine million US dollars in damages to agriculture, housing, and infrastructure. Rain rates associated with the

[7] Meteorological service database.
[8] Office of disaster Planning and Emergency Management (OFPEM) and Meteorological service database.
[9] Meteorological service database.

event were in excess of 30mm/hr and resulted in 425 mm of rainfall in Port Antonio (twice the normal January total), and a 100-year flood peak discharge along the Boundbrook River.

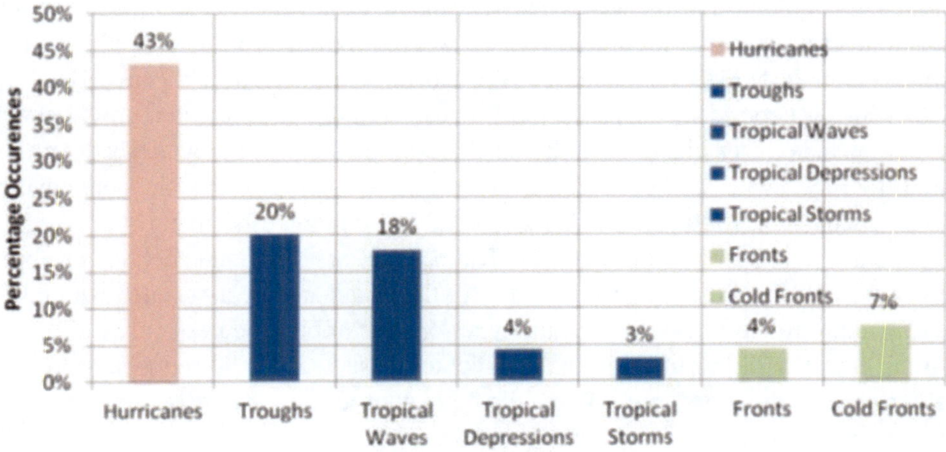

Figure 10.3. Occurrences of Meteorological phenomena in Jamaica between 1678 and 2011 that have resulted in significant flooding.

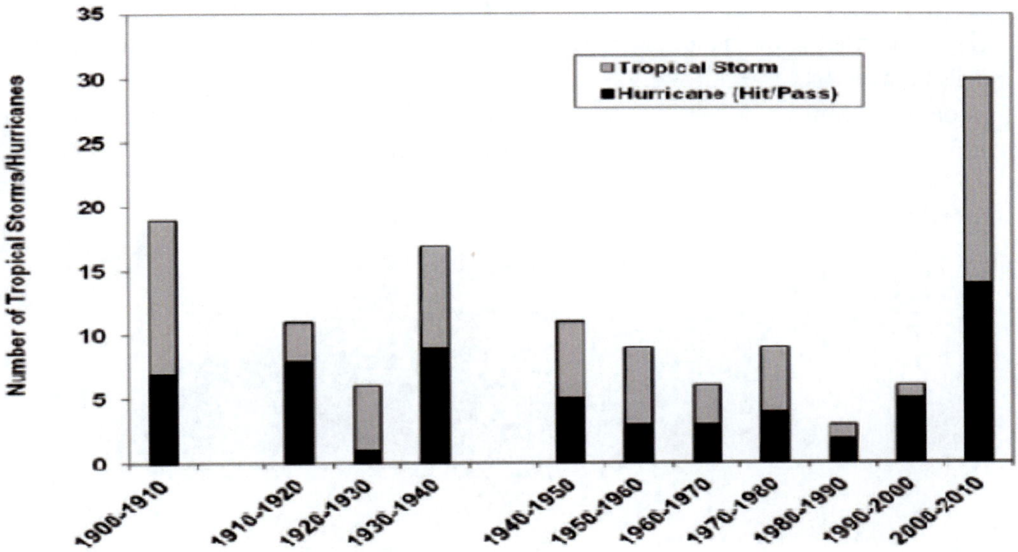

Data source: Meteorological Service of Jamaica.

Figure 10.4. The number of tropical storms /hurricanes which have hit or passed Jamaica from 1900-2010.

Severe flood events usually have a record of representative precipitation depth over the duration of the event. Total precipitation depth may range from 101 mm that flooded St. James in 1908 to 1030 mm in February 2002 that affected a number of parishes from St. Elizabeth in the southwest to St. Thomas in the east over an eleven day period. Analysis of severe events for the 1940 to 2010 period suggests that severe flooding is initiated at the 100 to 200 mm depth with a mean of 500 mm.

The link between rainfall and flooding is also evident in the spatial distribution of both variables across the island.

There is significant variability in the rainfall distribution pattern across the island with the north-east i.e., Portland and St. Mary, receiving a maximum in excess of 6000 mm/yr as opposed to the dry south-east which receives a low of ~700 mm/yr, as determined from the 30 year annual rainfall data for the island. Superimposing the location of reported flood events since 1904 and watershed locations on the 30 year rainfall mean distribution (Figure 10.5) shows the coincidence of higher rainfall and flood-prone areas in the watersheds of the Rio Grande, Buff Bay, Swift and Pencar rivers in the east, as well as in the Rio Cobre and Hope River watersheds in St. Catherine and Kingston and St. Andrew respectively.

In general the parishes of Portland, St. Thomas and St. Mary and sections of upper St. Andrew in the east lie in the higher rain zones for the island and are associated with rainfall – induced flooding from tropical storms, hurricanes or even tropical depressions.

Eastern Jamaica is particularly vulnerable to these weather systems as their tracks generally pass over the island in a NW-SE direction after leaving the Lesser Antilles and after moving across the warmer waters of the Atlantic.

3.2. Flooding, Topography and Lithology

Though rainfall is the major contributor to flooding, the topography and lithology of the island also play a role.

Flooding in Jamaica is of five types: Riverine, Groundwater, Storm Surge, Coastal and Urban flooding. Figure 10.6 below shows regions prone to the different types of flooding.

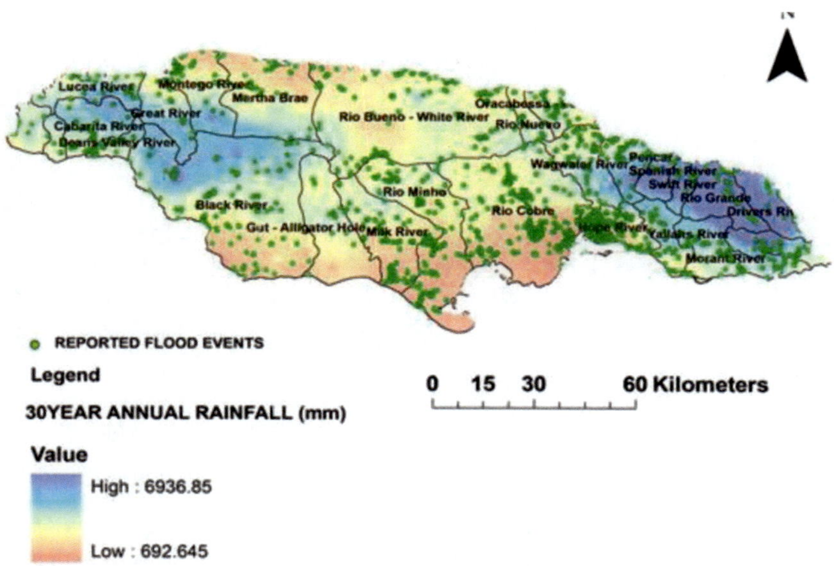

Mona Geoinfomatix Ltd (MGI).

Figure 10.5. Map of Jamaica showing the watersheds and flood-prone areas with relation to the rainfall pattern.

Riverine flooding is the main type of flooding in the eastern parishes of Portland, St. Thomas, sections of St. Mary and upper St. Andrew. Apart from lying in the high rain zones (see previous section), a higher incidence of surface water drainage and/or the presence of rivers accounts for the higher flood frequency in these regions. Flooding events are usually associated with high discharge of the major rivers e.g., the Rio Grande, Rio Cobre and Rio Minho. Nkemdirim (1978) showed that the Rio Grande and the Rio Cobre had a higher return period of flood peak flows as compared to the Roaring or Martha Brae rivers located in the western part of the island. This is observed since the Roaring and Martha Brae rivers drain the limestone aquifer while the Rio Grande and Rio Cobre are runoff from the impermeable volcanic rocks-flashy rivers On occasions, however, riverine flooding is associated with smaller rivers e.g., river discharge from the Port Maria or Outram river on November 23, 2006 due to a single 24 hour rainfall period of 240 mm caused massive flooding in the parish of St. Mary. Flood waters associated with the event ranged from 0.5 to 2 m in parts of the entire town causing damage to infrastructure. While the return period of the rainfall recorded at the Port Maria station is between 25 and 50 years, rain over Brimmer Hall also in St. Mary has a return period in excess of 100 years.

A better picture of the riverine flooding is shown in Figure 7 where the rivers present in the island have been superimposed on the major lithology or rock type of the island.

Occurrence of rivers or surface drainage versus groundwater flow depends on sub-surface lithology or rock type. Rock types are classified based on their properties to store or extract water (porosity and permeability) as aquifers and aquicludes. Those that have the ability to store or extract water are aquifers while those impermeable media which do not allow infiltration of water are aquicludes.

It is seen from Figure 10.7 that rivers exist in the regions where there is the presence of basal aquiclude comprised of volcanic, volcanoclastics and conglomerates. This includes the Blue Mountain region in the parish of Portland, the Richmond Formation in St. Mary and the Central Inlier located in the upper parts of Manchester and Clarendon.

When compared with an elevation model of the island it is noted that the rivers originate in the highlands of the Blue Mountains in the east and the Central Inlier and flow towards the coastlines following the slope of the land. High elevation coupled with high rainfall and the presence of an impervious surface accounts for the presence of riverine flooding in these areas.

Groundwater and depression flooding, on the other hand, is confined to the lowlands of St. Ann, Trelawny, sections of St. Catherine, Manchester and sections of Hanover where karstified limestones accounts for the major lithology.

Much of the island is comprised of tertiary White Limestones with high infiltration capacity and which account for the limestone aquifers of the island. Groundwater flooding in Manchester is confined to the areas of Porus and Harmons which are located in a depression or structural low called a graben, and which have a shallow water level which seeps up to the surface during increased rainfall.

Significant flooding occurred from the rains of May 22-June 13, 2002 which led to intensified aquifer recharge with a resulting increase in the groundwater table of between 18 and 98 M in the Upper Manchester area (WRA 2002).

Rainfall from tropical storms Isidore and Lili (September 2002) also resulted in the flooding of Porus (2.97 m increase of water level) and Harmons (6.80 m rise in water level)

(WRA 2002). Recent rains from Tropical storm Nicole (2010) also caused severe groundwater and depression flooding in Bliss Pasture (Trelawny) and Moneague (St. Ann).

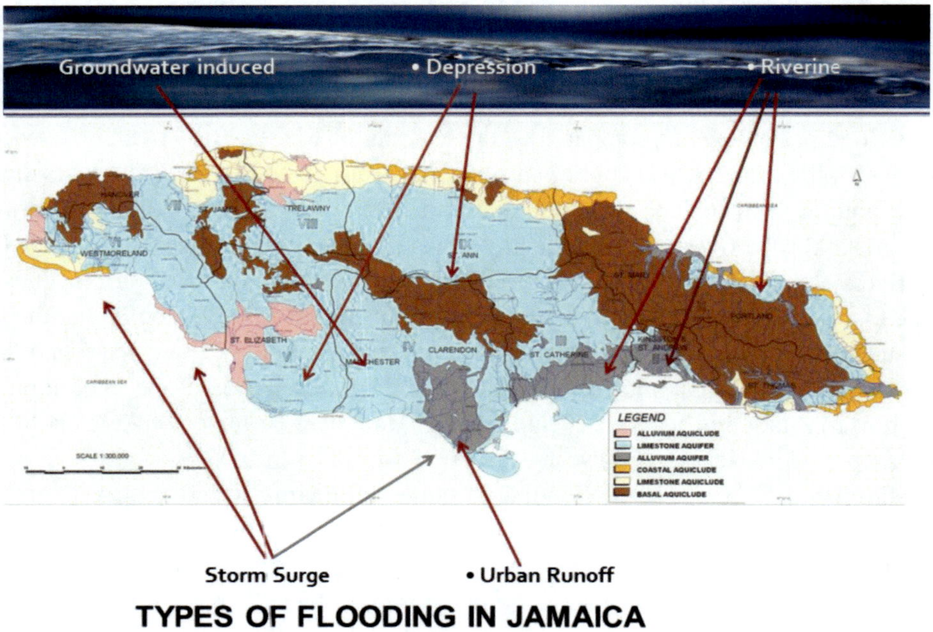

WRA, Jamaica.

Figure 10.6. Types of Flooding in Jamaica.

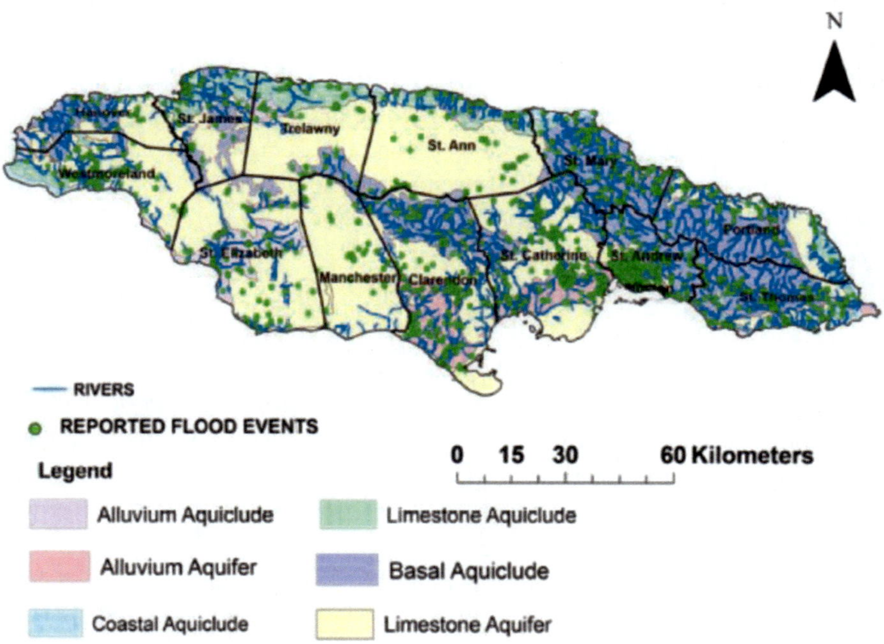

Source: WRA.

Figure 10.7. Map of Jamaica showing the different hydrostratigraphic units, rivers and location of flooding events.

Flooding in the Liguanea alluvium plains of Kingston located in the Hope river watershed is a combination of both urban and riverine flooding. High discharge from the upper watershed of St Andrew brings down high volumes of water along with debris and boulders causing flooding in the lower reaches of the Hope River at Kintyre and Harbour View. Floods in the Hope river watershed have caused repeated damage to a fording at Kintyre and the bridge at Harbour View. The latter collapsed during tropical storm Nicole (September 2010) disrupting commutation between the parishes of Kingston and St. Thomas. The bridge has since been reconstructed. The rains of September 30, 2012 followed by Hurricane Sandy (October 2012) caused the fording at Kintyre to collapse.

The pictures of Figure 10.8 depict the damage caused from flooding associated with the above mentioned events.

Coastal Flooding has been experienced around the entire coastline of Jamaica. This is frequently due to storm surge. Typically, extreme storm surge levels vary between 4.5 metres (Port Royal) and 4.2 metres (Old Harbour Bay in Hurricane Ivan) and are observed during the passage of tropical storms and hurricanes. These levels have historically inundated towns (Annotto Bay, Port Royal and Portland Cottage). Other factors that may contribute to coastal flooding include tsunamis, coastal subsidence and sea level rise. A tsunami is a series of water waves generated by earthquakes, volcanic processes, landslides, meteorite impacts or other remote sources. Tsunamis affecting Jamaica are largely generated from landslides.

Some of the earlier events that have impacted Jamaica have been recorded for Port Royal (March 1688 and June 1692), Savanna la mar (October 1780) and Annotto Bay (November 1812 and January 1907).

Coastal flooding also arises as a result of the increasing sea level s. During the 20th century, global sea level rose at an average rate of 4.8 to 8.8 inches per century (1.2-2.2 mm/ year) (Christensen et al., 2007). Estimates for the Caribbean suggest that sea level rise between 1950 and 2000 appears to be near the global mean (Church et al., 2004).

More recent examinations of satellite measurements however estimate that global sea level has been rising at an even more alarming rate of 9 to 15 inches per century (2.4-3.8 mm/yr) since 1993 (Bindoff et al., 2007). Sea levels measured at Port Royal between 1955 and 1971 also suggest an increasing trend of 0.9 mm/year (Horsfield 1973). Higher sea levels give rise to greater extents of coastal flooding since the base sea level on which the storm surge height is superimposed is raised (CARIBSAVE 2011).

3.3. Flooding, Climate Variability and Climate Change

Principal Component Analysis reveals two dominant patterns in annual rainfall over Jamaica (Figure 10.9). The pattern of Figure 10.9A explains 72.3% of rainfall variability and suggests that in the mean the entire island could be considered as a single rainfall zone which exhibits strong interannual (year-to-year) variability (Figure 10.9C).

Much of the interannual variability may be attributed to the El Niño -Southern Oscillation (ENSO) phenomenon (see for e.g., Chen and Taylor 2002) which varies on timescales of 3-7 years and which significantly alters rainfall in the Caribbean region dependent on the time of year and the phase it is in. For example, El Niño (La Niña) years tend to be characterized by fewer (more) tropical storms and hurricanes and generally a less (more) conducive environment for convection during the late wet season.

A. Collapse of Bridge at Harbour View, Kingston, September 2011, Tropical Storm Nicole. B. Collapse of fording at Kintyre, Kingston, Tropical Storm Nicole, September 2010. C. New bridge at Harbour View, undamaged by storms, hurricanes since 2010. D. Fording at Kintyre reconstructed as of April 2012. E. Fording at Kintyre and the low flow of the Hope River in April, 2012. F. Fording at Kintyre after Hurricane Sandy, October 26, 2012.

Figure 10.8. Impact of recent flooding events on infrastructure.

During the dry season of El Niño years, mid-latitude systems tend to pass well south of their normal tracks and fronts, prefrontal troughs, and upper-level low pressure systems can contribute to the development of deep convection and heavy precipitation in the north Caribbean (Laing 2004).

During the early rainfall season in years of El Niño decline rainfall is enhanced across the Caribbean (Chen and Taylor 2002) primarily due to a warming of the north tropical Atlantic. Not surprisingly, significant flood events have been linked to El Niño or La Niña occurrences. The years of highest reported flood events in the recent past (2007 and 2010) both coincide with La Niña events. Laing (2004) also notes three cases of heavy precipitation and flash floods in the north Caribbean, all of which occurred during the dry season of El Niño years: January 3-4, 1998 and March 4 1998 (Jamaica); January 5-6, 1992 (Puerto Rico).

The pattern of Figure 10.9B explains 20.7% of annual rainfall distribution and is largely representative of rainfall over eastern Jamaica.

Figure 10.9D indicates that this mode is characterized by decadal variation i.e., by groups of years when the country is background conditioned to be wet (e.g., years centered on 1985) or dry (1995 and beyond). The post 1995 behaviour of this mode is strongly reminiscent of both the upswing in tropical storm/ hurricane activity and flood events alluded to earlier, and is probably associated with the phase of the Atlantic Multi-decadal Oscillation (AMO).

The AMO is a multidecadal oscillation in sea surfaces temperatures over the tropical Atlantic which swung into a warm phase in the late 1990s and which has been shown to be linked to hurricane activity in the tropical Atlantic (Chylek and Lesins, 2008).

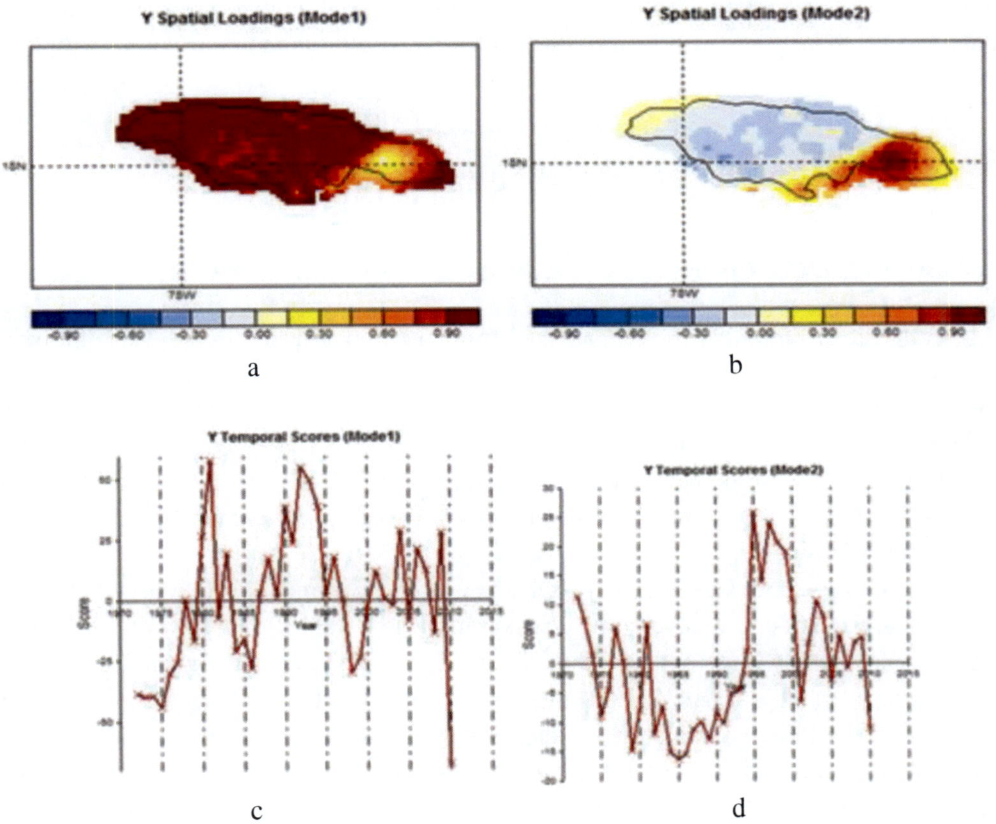

Figure 10.9. Two dominant patterns of rainfall distribution over Jamaica deduced from annual values using principal component analysis. The patterns shown in (A) and (B) explain 72.3% and 20.9% respectively of the overall annual rainfall pattern over Jamaica. Plots (C) and (D) shows how strongly patterns (A) and (B) respectively manifest each year since 1971.

Given that rainfall is the dominant factor in flood occurrences in Jamaica there is clear indication that flooding and flood conditions in Jamaica have global influencing factors, including ENSO and the AMO, which account for the cyclical patterns evident in the records. Given the ability to predict some of the global phenomenon in advance, this may bode well for planning and preparation for high flood event years.

In much the same way, climate change likely has and will continue to influence the frequency and/or intensity of flood events on the longer time scale. By the end of the current century, Jamaica is projected to be significantly dryer, with largest projected drying occurring during the summer rainfall season (CSGM 2012).

On the face of it this may suggest a reduction in flood events under a globally warmed world. The projections however also suggest a change in the character of rain events i.e., it is projected that heavy rainfall events may increase in the Caribbean even as the number of consecutive dry days also increases (McLean et al., 2011). The intensity of tropical cyclones is also projected to increase, though there is uncertainty about the impact on the frequency of storms. The suggestion is that though there may be uncertainty about the frequency of future flood events, the risk of flooding when rain does occur may be greater due to the higher intensities of rainfall. It is also noted that the future might be one of more El Niño events (Christensen et al., 2007) which might increase flood frequency during the dry season. In addition, the combination of more intense hurricanes and higher projected sea level s (CARIBSAVE 2011) may also lead to an increase in coastal inundation due to storm surge flooding.

4. Coping Strategies

4.1. Flooding, Development and the Regulatory Framework

In addition to rainfall received and the physical characteristics of the land, the risk of flooding at a given location is also a function of the adequacy of the natural channels and manmade drainage infrastructure and can be heightened or lessened by anthropogenic actions which impact the drainage systems.

Collymore [2007] reported several occurrences of repeated flooding in Jamaica (1979, 1984, 1985, 1986, 1988) as a result of the disharmony between human use of the environment and natural systems.

Within Jamaica, there are four key agencies which play a monitoring role with respect to flooding and infrastructural development which may alter water flow. They are the National Works Agency (NWA), the Water Resources Authority (WRA), the National Environment and Planning Agency (NEPA) and the Office of Disaster Planning and Emergency Management (ODPEM). Developments in Jamaica are subjected to approval from all four agencies as it relates to flood control. Development plans submitted to local Parish Councils are referred first to NEPA, which is a central planning authority, which in turn disseminates the drainage plans and reports to the other three agencies for review and comments.

General requirements which must be met include: (i) major drainage systems must accommodate the 25 year event (ii) bridge openings must accommodate the 50 year return period event, and (iii) developments must not to be sited in the 100 year flood plain.

Hydraulic reports are required for proposed sub-divisions and land development projects as stipulated by the Developers and Investment Manual (2007), Volume 3 Section 1[10].

The division of responsibilities amongst the agencies is as follows:

- The NWA is the executing agency of the Ministry of Transport, Works and Housing and is responsible for planning, building and maintaining drainage and flood control systems. The NWA undertakes river training and flood control projects and provides guidance for infrastructural developments through their assessments of on-site and off-site drainage. The NWA also declares Floodwater control areas (in conjunction with ODPEM), and prepares the Floodwater Control Master Plan.
- The WRA (whose parent ministry is the Ministry of Water, Land, Environment and Climate Change) is responsible for the management, protection, and controlled allocation and use of Jamaica's water resources. The WRA provides data, information, and technical assistance to both government and non-government institutions, and prepares flood plain maps. The WRA also reviews flood impact assessment s submitted to NEPA as a part of the development approval processes. They are a repository of flood related information as they have mapped flood-prone areas across Jamaica and have undertaken the flood plain studies of over nine major rivers across Jamaica up to 2010.
- ODPEM is a statutory body that operates out of the Office of the Prime Minister. ODPEM is responsible for conducting hazard identification and risk assessment s. Proposed developments in flood-prone areas are required to follow the guidelines of ODPEM.
- Like the NWA, NEPA is an Executive Agency which merges the Natural Resources Conservation Authority (NRCA), the Town Planning Department (TPD) and the Land Development and Utilization Commission (LDUC). The Agency integrates environmental, planning and sustainable development policies and programmes, by carrying out the legislative mandate of the Natural Resources Conservation Authority (NRCA) Act; The Town and Country Planning Act; and The Land Development and Utilization Act. The boards under each of these authorities regulate developments at a national level and inform the relevant local parish councils. NEPA provides guidelines to developers which include a requirement to have a drainage plan for the proposed development. NEPA also seeks advice from the NWA, WRA and ODPEM on the suitability of development plans.

4.2. Flooding and Modelling

There is a history of using hydrological modelling in Jamaica, particularly, with respect to large and newer infrastructural developments in designated flood-prone areas, but more recently with an emphasis on determining the vulnerabilities of communities located in the major watersheds to flooding under present and future climatic conditions. In general, with respect to infrastructural development, the hydrology of relatively small urban catchments is analysed using the Rational Method, while larger urban and rural catchments are analysed

[10] Developer's and Investment Manual (2007), Office of the Prime Minster.

with the Soil Conservation Service (SCS) method. Very large catchments of river systems are sometimes analysed with a Regional Flood Frequency method using gauged station data. Jamaica has such a regional method specifically defined for its catchments, as captured in the UWA (1987)[11]. The hydrological models (i.e., the Rational and SCS Methods) essentially define the relationship between precipitation and effective runoff.

The relationship between storm duration (usually of the order to 5 minutes to 5 days), storm intensity (usually expressed in mm per hour), and storm return interval (2 to 100 year return period) is represented by intensity-duration-frequency (IDF) curves derived from the analysis of storms at a given location. Due to data limitations, the current practice is to use the IDF curves developed for the two international airports. This represents a limitation to the application of the methodology.

Several methods exist for the estimation of rainfall excess or runoff hydrographs from direct 24-hour rainfall depths. The methods have wide applicability across rural Jamaica where over 200 24-hour stations/gauges exist. The SCS Method is the most popular because it is relatively simple to use and accounts for land use, soil type and antecedent conditions. The disadvantages of the method include the sensitivity of the runoff prediction to Curve Number values and the need to fix the initial abstraction ratio (to 0.2). The SCS method was originally developed by the United States Department of Agriculture for agricultural watersheds (USDA 1986). The current practice in Jamaica is to utilize the implementation of the SCS method either in spreadsheet implementations of the numerical model or in the HEC-HMS TR-55 model. Standard temporal rainfall patterns/distributions are usually applied to the abstraction (determined from land use and soil type) and storage component (derived from topography and soil type) of the SCS model to derive an effective runoff. The USDA has developed a set of standard design rainfall curves for different rainfall climates /regions across the continental US. There are four synthetic 24-hour rainfall distributions (I, IA, II, and III). Type IA is the least intense and type II represents the most intense short duration rainfall. Jamaican infrastructure development projects typically use Type II or III rainfall curves since Type III is applicable to the Gulf of Mexico and Atlantic coastal US where hurricanes and tropical storms bring large 24-hour rainfall amounts, while Type II represents the rest of the United States country. This, however, represents a gap in the hydrological knowledge of the country that needs to be addressed.

As a case study of the use of a hydrological model for predictive purposes and for determining vulnerabilities, the University of the West Indies (Mona), Department of Geography and Geology is involved in modelling the Hope River Watershed using the HEC HMS hydrological model.

HEC-HMS facilitates the computation of runoff from rainfall using the SCS curve number method, after a prior determination of the watershed boundaries, catchments, drainage lines and associated watershed characteristics. It uses a GIS based interface (HEC GEO-HMS) for determining the flow direction grid and flow accumulation grids. The river length (19km) is computed along with the upstream and downstream elevations of the river.

The catchments of the watersheds are delineated and information about their area, elevation and flow center captured (Figure 10.10a). Capturing the catchments is necessary in order to delineate the sub-basins and the outlet points of the main river. Soil and land use data are used to delineate areas susceptible to and affected by flooding.

[11] Underground water Authority (1987) regional flood frequency Analysis (H. Thomas).

For the Hope River watershed, the land use distribution map shows that the main important land use feature i.e., "buildings and infrastructures" are located within the Hydrologic soil Group A which is susceptible to high run-off due to less internal drainage (USDA 1972, 1986). The soil and land use along with the other parameters are merged in GEO-HMS to create the curve number. The curve number is then used as a standard value by the SCS (Soil Conservation System) to calculate the run-off using empirical methods.

Though the method can be used for ungauged basins, in the case of the Hope River (which is gauged), the discharge is calculated in HEC-HMS from rainfall data utilising the SCS CN (Curve Number) lag method. That is, the discharge is calculated by computing unit hydrographs which show the peak value for concentration (runoff) and the lag time taken for the water to reach the peak value in the basin. For the project, rainfall data (from 1992-2011) was obtained from the Meteorological Service of Jamaica for the stations shown in Figure 10.10a. Streamflow analysis of the Hope River near Gordon Town, Penfield and of the Mammee River at Gordon town provided daily measured gauge data from 1955 (Hope River) and from 1988 (Mammee and Hope at Penfield) to 2010. The instantaneous peaks (Figure 10.11) as obtained from the data set show peak discharges associated with some of the significant hurricanes and tropical storms to impact Jamaica in the recent past.

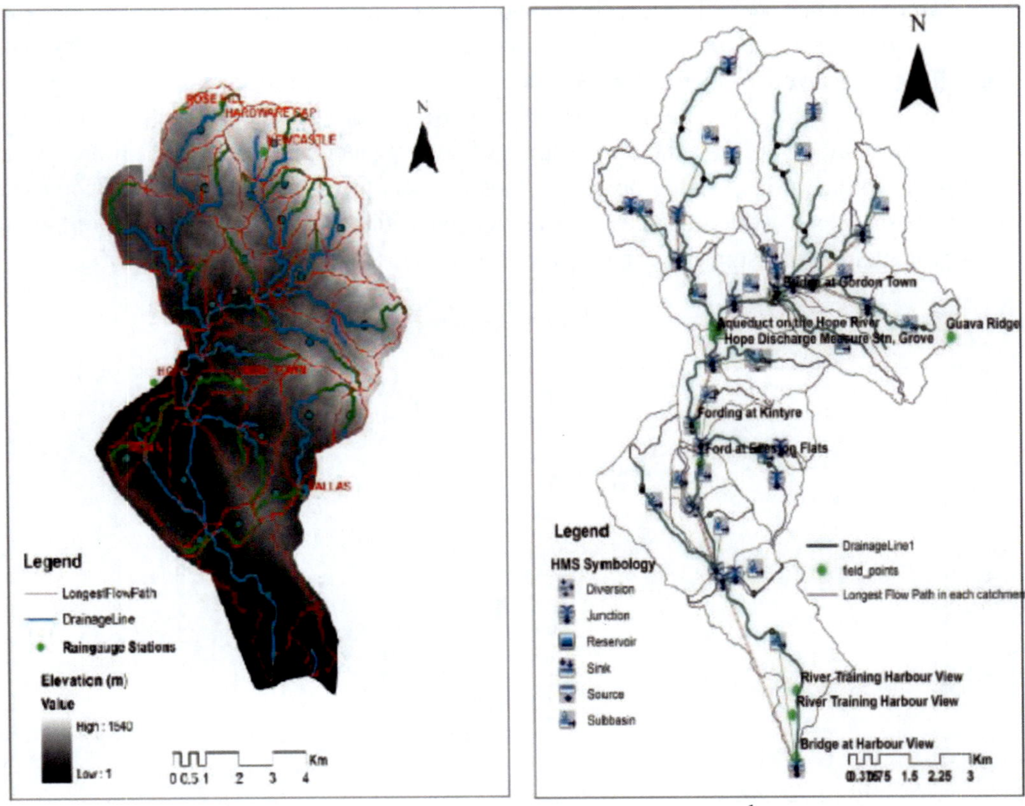

a b

Figure 10.10. (a) Digital elevation map of the watershed showing the drainage line, maximum flow path for the catchments, centroid of the catchments and the raingauge stations in the watershed. (b) Basin Model for the Hope River Watershed.

A digital elevation model (DEM) of 5m horizontal and 1m vertical resolution was used to develop the basin model for the watershed in HEC-GEOHMS. The resultant basin model shows all the attributes of the watershed, i.e., the length of the longest flow paths, drainage lines, upstream and downstream slope of the river and the locations of the different sub-basins (catchments), junctions of flow from two sub-basins and respective outlets of flow from each sub-basin of the watershed (Figure 10.10b).

As a preliminary run, the model was run using 5 minute rainfall intensity data for tropical storm Nicole measured at the Hope discharge station at Grove which is monitored by the Water Resources Authority of Jamaica. The model was also run using 5 minute rainfall data from June 1-11, 2011 and calibrated with discharge data obtained from the stream gauge at the Hope discharge station defined by Outlet 62 in the basin model, as determined in HEC GEO-HMS. The SCS curve numbers for each sub-basins which were determined in HEC-GEOHMS using soil and land use data and the basin lag time were used as other input data. Sample results of the runoff generated from the rainfall of Nicole at the two main locations of damage i.e., the fording at Kintyre as well as at Outlet 62 (of basin model) which corresponds to the location of the Hope discharge station at Grove are depicted in Figure 10.12 a and b.

The model can be run with different rainfall events either extreme or using the 100 year return period rainfall values for different stations. In the latter instance the Theissen Polygon method which assigns weights to each polygon based on the percentage of the area they occupy with respect to the total area of the watershed is applied. The usefulness of this kind of modelling is derived from the fact that the resultant output runoff can be simulated at different sections of the watersheds from a particular rainfall event. This kind of information is valuable for estimating the time and volume of water generated from each rainfall event. Accordingly, communities located nearby to these outlets or junctions can be provided with an ensemble of possible flooding scenarios associated with different extremes of rainfall and necessary adaptation measures can be taken to protect lives and livelihoods.

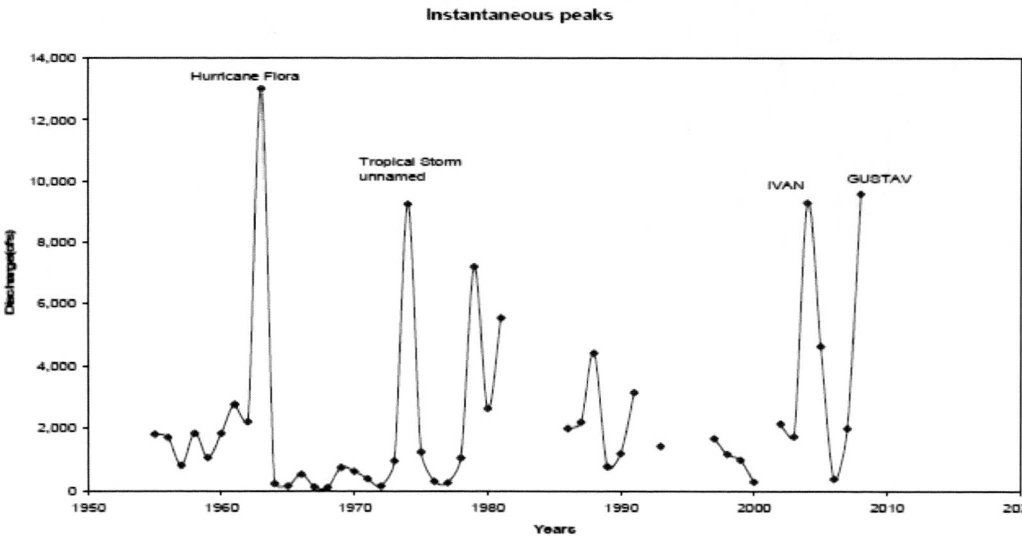

Data Source: Water Resources Authority of Jamaica.

Figure 10.11. Instantaneous peaks of the streamflow data at the Hope Discharge station at Gordon Town.

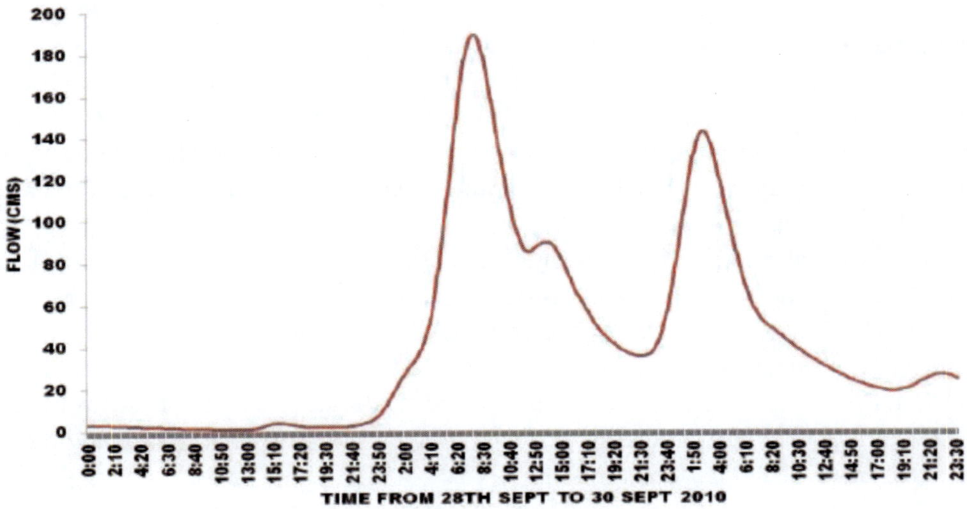

a

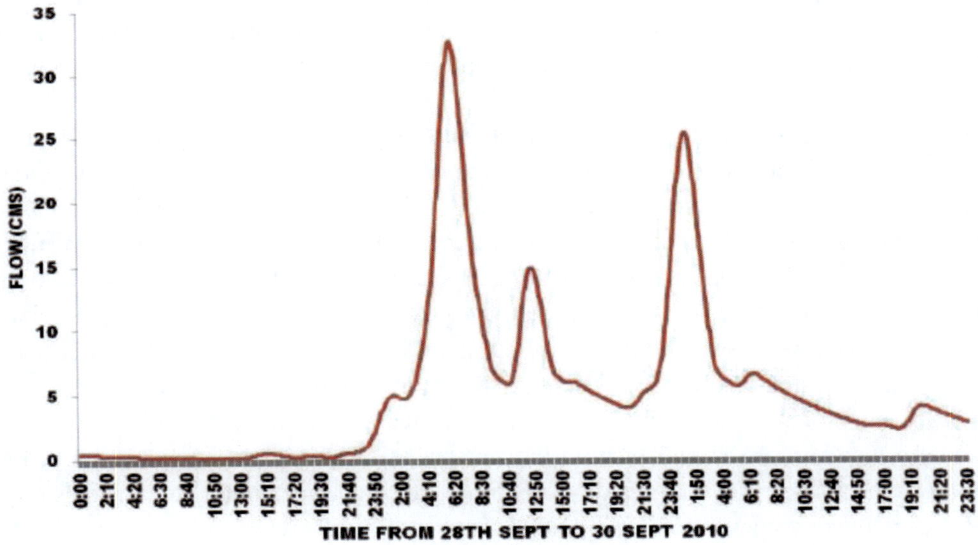

b

Figure 10.12. Simulated peak discharge from the model at (a) Outlet 62 corresponding to the locations of Hope discharge station at Grove and at (b) Junction 428 as corresponding to the fording at Kintyre, Kingston.

4.3. Flooding and Flood Hazard Management

The practice of flood hazard management is increasingly being recognized as critical for Jamaica. The practice of flood management and control in Jamaica has evolved in the recent past and utilizes some or all of the following to varying degrees and at various levels: Public Awareness Campaigns, Government Regulations, calculation of Floor levels and Return Periods, Infrastructural Solutions (detention Basins, Drain Improvements, Gully Restoration, Bridge Building), the strengthening of Community and governmental Institutional Capacity,

Risk Assessments, Flood Mapping and the Implementation of Flood Warning Systems. Some emphasis has been placed on the latter two methodologies.

Flood warning systems are an attractive solution given the relatively low cost of operation (being a non-structural approach) and the fact that they complement well the operation of other flood mitigation methods. Jamaica operates two kinds of flood warning systems: (i) a community operated flood warning system, and (ii) an automatic (real time) flood warning system.

In both cases the aim is to ensure that the end user receives appropriate warnings in time to take effective action to save lives and minimize damage to property. In the case of the community flood warning systems, observers make keen note of rainfall and water level gauges and relay information to response teams and the disaster agencies when alert and critical flood levels are reached. The community then decides whether or not to evacuate. With the automatic flood warning system rainfall and river gauges are installed at critical points in the watershed or its sub-basins.

Data from the different gauges in different watersheds are transmitted via telemetry to the central station located in the Headquarters of the WRA in Kingston for processing. When flooding is likely to occur, a flood warning will be issued by the Office of Disaster Planning and Emergency Management (ODPEM) and appropriate action taken to reduce damage and save lives.

Flood warning systems have been implemented in several communities throughout Jamaica with some successes including the saving of lives, livestock and personal belongings in Cave River, Rio Cobre and Rio Grande Valleys. Note that critical to either system is a community organizational structure and a willingness of community members to participate and collaborate with government. Government partners include the ODPEM who train community flood response teams and coordinate post-disaster activities; the Meteorological Service of Jamaica who issue the forecasts and monitor rain events; and the Water Resources Authority who design the early warning systems, train community flood reserve teams, installs and monitor the automatic systems and set alert and critical flood levels.

Finally, flood hazard management in Jamaica has also been enhanced by the use of Flood Plain Maps in Jamaica. The flood plain maps show the extent of flooding and the elevation of flood waters above sea level corresponding to flows of different return periods e.g., 10, 25, 50 and 100 years. Flood Plain maps exist for a number of the major rivers in Jamaica including the Rio Minho, Rio Grande, Rio Cobre, Yallahs and Hope Rivers and are used as the basis for flood plain management. Table 10.2, adapted from a 2003 WRA document, shows the primary users of flood plain maps in Jamaica and indicates their usefulness for planning and preparedness.

In general, the WRA (2003) notes that successfully tackling flood hazard management in Jamaica means tackling issues related to land use planning, flood plain regulations, flood water control, preservation of flood ways, channel rights of way and clearances, use of flood warning systems, the quantification of economic benefits, and the role and cost of insurance. They also note that that the primary challenges to flood hazard management in Jamaica include (i) defining clear roles and responsibilities for relevant agencies, (ii) implementing a coordinated approach to both flood plain mapping and early warning systems and (iii) defining land use options including designating floodways, flood zones and evacuation routes. That is, it is an integrated approach by forecasters, emergency managers, hydrologists, geologists, public works departments, and community residents that will reduce flood risk.

Table 10.2. Primary users of Flood Plain Maps in Jamaica

	ORGANIZATION	USE
1.	National Environment and Planning Agency	Locating housing developments, land use zoning
2.	Local Authorities	Locating housing developments
3.	Office of Disaster Preparedness and Emergency Management	Identification of vulnerable communities
4.	Private developers	Locating housing developments
5.	Ministry of Construction and Works	Design of roadways and bridges
6.	Mines and Geology Division	Licensing surface mining activities
7.	Tourism Product Development Company	Permits for river rafting
8.	Banana Estates	Determine land use

Adapted from WRA 2003.

CONCLUSION

Jamaica is exposed to multiple hazards including high risk for hurricanes, earthquakes and flooding. The experience of Jamaica illustrates that a good understanding of the characteristics of each hazard is important for planning purposes, particularly in the small island context, and given the cost and set back to the economy associated with the impact of the hazard. For Jamaica, a good understanding of flooding and flooding potential is necessary. For example, infrastructural design of many public-good structures (e.g., bridges and culverts) is very dependent on knowledge of the magnitude and duration of expected floods at the location. There is a growing urgency to acquire the necessary understanding given the context of yearly flooding and an upward trend in the frequency of occurrence, and the likelihood of further increased vulnerability of the island to flood events under climate change. There are several examples in Jamaica's recent past which illustrate that where adequate considerations were not given the resulting damage was great and costly.

There is clear need, then, for awareness of both the nature of and potential for flooding on the part of planners and of residents living in vulnerable flood-prone areas. Because flooding in Jamaica is strongly tied to the rainfall season it in one sense bodes well for planning and preparedness considerations. However the severity and impact of a flooding event is not solely linked to the rain-bearing event as other factors can be key determinants of the occurrence and extent of flooding including lithology (which also helps to define flood-prone areas) and non-natural factors such as land use. This suggests that coping in the future will require the implementation of a flood hazard management strategy, and the adoption of the mindset that such a strategy is an important tool for economic and development planning.

It is reiterated that it is an integrated approach that will reduce flood risk as the weakest link in the planning-preparedness -mitigation-detection -forecasting -warning-response process will jeopardize the entire effort.

ACKNOWLEDGMENTS

The following are recognized for help in compiling this chapter. The Water Resources Authority, and in particular Mr. Basil Fernandez, Mona Geoinformatix Institute Limited of the University Of The West Indies, Mona, Jamaica, the Meteorological Service of Jamaica, the Office of Disaster Preparedness and Emergency Management, Jamaica, Mr. Lawrence Barrett, and Dr. David Smith of the Institute for Sustainable Development at the University of the West Indies, Mona, Jamaica.

REFERENCES

Ahmad, R. and Carby, B. E. (1993). Vulnerability of Roads and Water Systems to Hydro-Geological Hazards in Jamaica. *Built Environment* v. 21 (2/3).

Bindoff, N. L., Willebrand, J., Artale, V., Cazenave, A., Gregory, J., Gulev, S., Hanawa, K., Le Quéré, C., Levitus, S., Noijiri, Y., Shum, C. K., Talley, L. D., and Unnikrishnan, A. (2007): Observations: oceanic climate change and sea level. In: Solomon, S., et al. (eds.), *Climate Change 2007: The Physical Science Basis*, Intergovernmental Panel on Climate Change. Cambridge: Cambridge University Press, pp. 385–432.

CARIBSAVE (2011). *The CARIBSAVE Climate Change Risk Atlas – Jamaica, Final draft Country Risk Profile*. CARIBSAVE

Church, J. A., White, N. J., Coleman, R., Lambeck, K., and Mitrovica, J. X., (2004). Estimates of the regional distribution of sea-level rise over the 1950 to 2000 period. *Journal of Climate*, 17(13), 2609– 2625.

Chen, A. A. and Taylor, M. A. (2002). Investigating the link between early season Caribbean rainfall and the El Niño +1 year. *Int. J. Climatol.*, 22, 87-106.

Christensen, J. H., Hewitson, B., Busuioc, A., Chen, A., Gao, A., Held, I., Jones, R., Kolli, R. K., Kwon, W.-T., Laprise, R., Magaña Rueda, V., Mearns, L., Menéndez, C. G., Räisänen, J., Rinke, A., Sarr, A., and Whetton, P. (2007). Regional Climate Projections. In: *Climate Change 2007: The Physical Science Basis. Contribution of Working Group I to the Fourth Assessment Report of the Intergovernmental Panel on Climate Change* [Solomon, S., D. Qin, M. Manning, Z. Chen, M. Marquis, K. B. Averyt, M. Tignor and H. L. Miller (eds.)]. Cambridge University Press, Cambridge, United Kingdom and New York, NY, US.

Chylek, P. and Lesins, G. (2008). Multidecadal variability of Atlantic hurricane activity: 1851–2007. *Journal of Geophysical Research* 113: D22106.

Climate Studies Group, Mona (CSGM), 2012: State of the Jamaica Climate: Past and Future. Information for Resilience Building. For Pilot Project for Climate Resilience. *GOJ.* 180 pp. ISBN 978 976 8103 41 3

Collymore, J. (2007). Disaster Impacts in the Caribbean. In: *International Perspectives on Natural Disasters: Occurrence, Mitigation and Consequence,* 303-322.

Developer's and Investment Manual (2007), Office of the Prime Minster.

Horsfield, W. T. (1973). Late Tertiary and Quaternary crustal movements in Jamaica. *Journal of the Geological Society of Jamaica*, 13: 6-13.

Laing, A. (2004). Cases of Heavy Precipitation and Flash Floods in the Caribbean during El Niño Winters. *J. Hydrometeor.*, 5, 577–594.

Langhammer, J. (2006). Geographic Approaches to Flood Risk Modelling. In: *Modelling Natural Environment and Society*, 81-99.

McLean, N., T. S. Stephenson and M. A. Taylor, 2011: Examining future Caribbean climate extremes from the ECHAM-driven PRECIS regional model, *WCRP Open Science Conference*, Denver, CO, US, October 24-28, 2011.

Nkemdirim, L. C. and Jones, E. B. (1978): Flood regions in Jamaica. *Hydrological Sciences Bulletin des Sciences Hydrologiques,* 23, 1, 3/1978.

ODPEM (2000) *Disaster Catalogue: Flood catalogue.*

Planning Institute of Jamaica (2009). *Vision 2030 Jamaica: national development plan.*

Planning Institute of Jamaica (2012). *Damage and Loss Assessment Reports*, http://pioj.gov.jm/ResearchandData/tabid/82/Default.aspx.

Planning Institute of Jamaica (2010). *Selected Indicators*, website: http://www.pioj.gov.jm/Portals/0/Economic_Sector/SELECTED%20INDICATORS%202005%20to%202010.pdf

Underground Water Authority (1987). *Regional Flood Frequency Analysis* (R. Thomas).

US Soil Conservation Service. 1986. *Urban Hydrology for Small Watersheds.* USDA (US Department of Agriculture). *Technical Release.*

Water Resources Authority, Jamaica (WRA) (2003): *Flood Hazard Management, Mapping and Early Warning Systems In Jamaica.* Author Errol Douglas. www.unisdr.org/2006/ppew/info-resources /.../Errol_Douglas.doc Accessed online: January 26, 2013.

In: Flooding and Climate Change
Editors: D. D. Chadee, J. M. Sutherland and J. B. Agard
ISBN: 978-1-62948-317-7

Chapter 11

FLOODING AND THE LAW

Camanie Naraynsingh-Chang*
Department of Para-Clinical Sciences, Faculty of Medicine,
The University of the West Indies, St. Augustine, Trinidad, West Indies

ABSTRACT

Flooding at the global, regional and local levels is introduced and the major causes of flooding in Trinidad and Tobago are listed. How Common Law has developed in Trinidad and Tobago is described. The major Statutes relating to flooding are also described. These are not found in a single statute but are found in various laws including the post 2000 Environmental Management act and its 2007 Amendment. The Certificate of Clearance (Designated Activities) Order 2001 outlines the activities that require a certificate of Environmental Clearance (CEC). Among the 44 activities listed are those that are closely linked to flooding. The Agriculture Fires Act, the Forest Act, the Malicious Damage Act and the Sawmills Act all Contain provisions which should control the flow of water. Despite these provisions there is difficulty in proving contravention cases. Weakness is identified in the Town and Country Planning Division that fails to prevent unauthorised settlements on hillsides. Municipal corporations frequently fail to keep drains clean or prevent dumping in watercourses. The Environmental Management Agency (EMA) has its own environmental police but failures to apply for Certificates of Environmental Clearance (CECs) are frequent. Comprehensive legislation relating to the causes of flooding such as the Flood Risk Management (Scotland) Act 2009 is recommended.

1. INTRODUCTION

There is ample evidence that flooding is increasing worldwide. Satellite studies show that major river deltas are sinking so increasing the risk of flooding. [1] *Nature and Geoscience*, [2] also links significant flooding to densely populated river deltas that are often used for farming. These studies indicate that 85% of the deltas have experienced serious flooding in

* Corresponding author: E-mail: camaniechang@gmail.com.

the last ten years. The area affected is 260,000km^2, more than 50 times the size of Trinidad and even larger than all of Guyana. It is expected that the surface area affected by flooding will increase by about 50% in the next 10 years.

In July 2010 Pakistan experienced one of its worst flooding disasters- more than 1,000 people died and more than 2.5 million were affected in some way. [3] China had experienced the worst floods in a decade, leaving more than 700 dead and millions homeless. [4]

Caribbean countries have not been immune to the effects of heavy rainfall and have faced increased flooding in recent years. In May 2004, in the Dominican Republic and Haiti more than 550 people were killed. The flood waters had cut through the land destroying village after village on its way.

In Guyana in January 2005, flood waters shut down the capital, Georgetown [5] and in Trinidad in November 2008, unusually heavy downpours led to the shutdown of City Gate, the transportation hub in the capital, Port of Spain. A bridge on the main highway had collapsed and traffic was gridlocked.

1.1. The Caribbean

The Caribbean consists of a chain of about 7000 islands, islets, cays and reefs with a population of approximately 40 million people [6]. The arc stretches for about 4000 km from Bahamas in the North to Guyana and Suriname on the South America n mainland.

In the North, the islands of Haiti, Dominican Republic, Jamaica, and Puerto Rico are mountainous, the soil of the islands of St. Vincent, Dominica and St. Kitts consist of volcanic ash and is very fertile. Guyana and Suriname have muddy sea coasts with large expanses below sea level.

Photo courtesy: Guyana Times

Figure 11.1. Flooding in Guyana, South America.

The smaller, flatter islands such as The Bahamas and Turks and Caicos Islands tend to be dry with little or no surface water. The more elevated and bigger islands are characterised by the presence of surface water and forested hills and/or mountains. [7] The elevation of the islands have implications for flooding - the flatter the island, the more likely it is to flood and the more rapid the runoff from the higher regions to the lower.

The size of the island also has implications relating to the coastline. Where there is a combination of high population density and a small island size, coastal development is inevitable. This means that the integrity of the coast line would be negatively affected with resulting coastal erosion.

The seasons in the Caribbean are simply the dry season and the wet season, the dry season being from December to May and the wet season from June to November. Guyana and Surinam are exceptions to this as they have two wet seasons and two dry seasons. The wet seasons are from early May to mid July and from late November to late January while the periods in between are referred to as the dry seasons.

The region contains some of the poorest countries in the world with low GDP and high debt repayment. The economies of most islands of the Caribbean are based in tourism and agriculture which gives the effects of climate change major significance. Any change in weather conditions will affect either or both of these industries, and flood waters have often wiped out entire crops and livestock.

Increasing population size means increased requirement for food production, increased housing and with development, increased industrialisation. For all these activities to take place substantial areas of land must be cleared and therein lies the vulnerability to floodwaters.

2. Global Response to Climate Change

In discussing climate change, there is a tendency to believe that the effects are in the distant future. The evidence of the increase in flooding, however, is the most immediate sign of changing weather patterns.

The United Nations Framework Convention on Climate Change (UNFCCC) has initiated several major moves to bring governments to the table to halt the damage to the environment. The intention is for governments to commit to a global programme through the legal framework in their own countries.

The UNFCCC made its first big step with an international agreement, the Kyoto Protocol, so named because it was adopted in Japan in 1997.

The major feature of the Kyoto Protocol is that it sets binding targets for 37 industrialised countries and the European Community for reducing greenhouse gas (GHG) emissions. These amounts to an average of five per cent against 1990 levels over the five-year period 2008-2012.

The primary distinction between the Protocol and the preceding Convention is that the Convention *encourages* industrialised countries to stabilise GHG emissions while the Protocol *commits* them to do so.

Under the Treaty, countries must meet their targets primarily through national measures, often through policy directives and legislation.

The developed countries have engaged a number of developing countries in the Kyoto Protocol through the carbon market. One such in the Caribbean is Guyana, one of the most flood prone countries in the region, which will preserve its forests in exchange for funding from the Norwegian government.

The Kyoto Protocol focuses on compliance and enforcement. As the term for this protocol ends (2012), the enforcement branch of the UNFCCC reports regularly on monitoring of the signatories and the degree of compliance.

2.1. Definition of Flooding

A flood has been variously defined as "an overflowing of a large amount of water beyond its normal confines, especially over what is normally dry land" [8]

Causes of Flooding

Flooding is normally caused by natural weather events such as:

- heavy rainfall and thunderstorms, mainly over a short period;
- rivers overflowing their banks ;
- a combination of storms, tides and low air pressure

The flooding is compounded by activities that propel the flow of water into low lying areas.

2.2. Trinidad and Tobago

The Environmental Management Agency (EMA) of Trinidad and Tobago reports that the twin island republic has been subjected to "unprecedented flooding events" in the past few years.

The EMA gives the following as causes of flooding [9]

- Indiscriminate clearing of land resulting in increased runoff, often times laden with silt and debris from the site, which eventually end up in our waterways ;
- Rapid development resulting in conversion of pervious surfaces into impervious surfaces. Often this is the main contributor to urban flooding ;
- Indiscriminate dumping of garbage ranging from plastic bottles and bags to old refrigerators and mattresses;
- Inadequate existing drainage networks;
- Construction on natural floodplains;
- Flood protection measures in upstream areas leading to increased flooding downstream

In the Caribbean where there has been a rash of development in recent years and new settlements mushroom all over the landscape, the causes of flooding are much the same. While there is significant coastal erosion in most of the islands, Guyana and Suriname are particularly vulnerable. Already below sea level they face even greater threats. In addition it is not unusual for those countries to experience severe flooding at spring tides, where waters breach the coastal defences, wreck polders and bring tremendous hardship to those living in affected areas.

Experts have projected that the sea level in both these countries will rise at a rate of one centimetre (cm) per year and about 40 to 60 cm by the end of the 21st century. A rise in sea level in the case of Guyana and Suriname without necessary protection means the inundation of wetlands and lowlands, acceleration of coastal erosion, flooding, the threatening or disruption of coastal structures, rage and water tables and increase in salinity of rivers and aquifers. [10]

3. The Law

3.1. Trinidad and Tobago - Case Study

The laws that govern flooding derive from either the Common law or Legislation passed in Parliament. Our laws in Trinidad and Tobago, as in other English speaking Caribbean countries, unless they are enacted in parliament, derive either from the Common Law in England or from Old Ordinances that were never changed.

3.2. Common Law

Decisions of superior courts are in themselves a source of law. In Trinidad and Tobago, decisions of the Privy Council are binding on all our courts. As the earlier courts made decisions, their decisions were recorded and cited in cases coming before the courts. Judges look to see whether the circumstances in which a decision was held is similar to the case before the court and whether the same should apply. These decisions replace old laws and customs. The Common Law refers to law that develops over time from judgments passed. It really is a condensation of jurisprudence over time and it develops as situations change.

Usually when one thinks of flooding, one envisions large acreages of land covered by water. This is more often the case. There are cases however where flooding occurs because of the action taken by a person in a neighbouring plot of land. In these circumstances, damages may be retrieved in a court of law using the common law doctrines of nuisance, negligence or the rule in *Rylands v Fletcher.*

Nuisance is regarded as the tort that deals with protection of the environment. A nuisance is committed when someone “does something on his own land which is not confined to his own land but extends to the land of his neighbour by

1. Causing an encroachment on his neighbour's land when it closely resembles trespass,
2. By causing physical damage to his neighbour's land or building or works or vegetation upon it, or
3. Unduly interfering with his neighbour in the comfortable and convenient enjoyment of his land" [11]

The law in private nuisance tries to balance the rights of an owner to enjoy his land with that of his neighbour in enjoyment of his land [12]. In Greencock Corporation v Canadian Railway [13] Lord Findlay stated "it is the duty of anyone who interferes with the course of a stream to see that the works which he substitutes for the channel provided by nature are adequate to carry off the water brought down even by extraordinary rainfall, and if damage results from the deficiency of the substitute which he has provided for the natural channel, he will be liable.

In *Arscott v The Coal Authority [2004]* [14] the "common enemy" principle was discussed. The case concerns a serious flood which took place at Aberfan, Merthyr Tydfil, on 22/23 October 1998. That night there was exceptionally heavy rainfall which caused the river to overflow its banks and the homes of the 32 appellants suffered substantial damage. The appellants blamed the Authority which had deposited colliery waste on a piece of land it owned causing the land to be elevated. One of the questions raised was whether the filling caused the properties to be flooded and it was determined that it did. The judge however relied on an 18th century principle known as the "common enemy rule". He articulated this rule as follows,"...an owner or occupier of land is entitled to use or develop his land so as to prevent flood waters coming on to his land. If in times of flood waters which would have entered his land in consequence damage another's land – that does not provide a cause of action in nuisance." He cited *R v The Commissioners of Sewers for the Levels of Pagham* [15]. The Commissioners erected groynes and other works to defend the stretch of coast for which they were responsible against the sea's encroachment. But the consequence was that the sea flowed with greater force upon adjoining land, whose owner brought proceedings. Lord Tenterden CJ said at 361:

> "I am... of opinion that the only safe rule to lay down is this, that each land-owner for himself, or the commissioners acting for several land-owners, may erect such defences for the land under their care as the necessity of the case requires, leaving it to others, in like manner, to protect themselves against the common enemy."

The limitations on these rules are that there must be no active interference with the river or watercourse itself and the defences to the watercourse must be on the landowner's land.

In *Rylands v Fletcher* [16] it was stated

> " that the person who for his own purposes brings on his land and collects and keeps there anything likely to do mischief if it escapes must keep it in at his peril, and if he does not do so, is prima facie answerable for all the damage which is a natural consequence of its escape."
>
> The rule in *Rylands v Fletcher* is thought to apply where there is unnatural user and where there is an escape.

The defence of *Act of God* applies "in circumstances which no human foresight can provide against and of which human prudence is not bound to recognise the possibility" [17]

In *Nichols v Marsland* [18] this defence found favour. The defendant had artificial ornamental lakes on his property which overflowed and caused damage after rainfall that was "greater and more violent than any within the memory of witnesses ". A jury found that the owner of the lakes was not negligent. Courts nowadays restrict this defence.

In Trinidad and Tobago cases follow the principles of English Law.

3.3. Legislation

The Statutes relating to flooding are not found in any single piece of legislation but are included in various laws of Trinidad and Tobago.

These in major part are:

Before 2000:

The Town and Country Planning Act Chap 35:02
The Municipal Corporations Act 21of 1990
The Agriculture Fires Act
The Malicious Damage Act
The Forest Act
The State Lands Act
The Litter Act

Post 2000

The Environmental Management Act
The Certificate of Environmental Clearance (Designated Activities) (Amendment) Order 2007
The Minerals Act

3.4. Town and Country Planning Act

One of the main pieces of legislation to deal with the environment prior to 2000 is the Town and Country Planning Act, Chap 35:01 which is an Act to make provision for orderly progressive development and control of land use throughout the country. "Development" means the carrying out of building, engineering, mining or other operations in, on, over or under any land, the making of any material change in the use of any buildings or other land, or the subdivision of any land.

The Minister has overall responsibility for formulation of a comprehensive policy for land use and development. The land use map must allocate areas for agricultural use, housing and industrial areas. The Act confers powers in respect of acquisition and development of land for planning. In addition the Act establishes an Advisory Town Planning Panel which advises the Minister on any matter within their knowledge or on any matter in which the Minister will seek their advice. There is no further mention of this panel in the Act except in the schedule that sets out the constitution of the panel.

The Minister may delegate responsibility to the local authority e.g., granting permission for land use or erection of buildings but he also has power to overturn a decision of the local authority.

The Minister might also ask that requests for planning permission of a particular nature be referred to him and has power to halt development which takes place without permission. The decision of the Minister is final.

An enforcement notice detailing the steps to be taken over a specified period of time in order to rectify the situation is served on a developer when he carries out work on the land without permission. Where enforcement notices are not complied with, the Minister may enter land and take the necessary steps.

But the Town and Country Planning Act does not deal with the environment in isolation. Because flooding derives from a variety of factors, then not unexpectedly various pieces of legislation are in place to deal with specific factors. The Municipal Corporations Act No. 21 of 1990, Timber Act, the Agricultural Fires Act, the State Lands Act and the Forest act all have their role to play in proper management of the environment.

3.5. The Municipal Corporations Act

Municipal corporations also have wide ranging responsibilities which include, for the purposes of this paper, the maintenance of watercourses and drains, and granting permission to erect buildings or making additions to buildings.

The Municipal Corporations Act No 21 of 1990 at section 145 makes it an offence to impede the free flow of water in any ditch, drain or water course within a municipality that is liable to a fine of one thousand dollars and to a further fine of one hundred dollars a day for each day that the offence continues after conviction.

Each corporation has a Municipal Police Service to deal with infractions of the law and the offence in the above paragraph is dealt with by these officers.

3.6. The Agricultural Fires Act

The EMA has listed fires as one of the major causes for rapid run off and flooding.

The Agricultural Fires Act provides that each County must have a County Fire Control Officer and identifies a period of time known as the fire season during which, if a person wishes to light a non-domestic fire, he must first obtain permission from a County Fire Control Officer. In addition, the Act identifies and sets out measures that must be taken to prevent fires in areas that are close to the forest.

Despite this piece of legislation, Trinidad and Tobago has seen significant fire damage over large tracts of land. In 2010 with its lengthy period of drought not unexpectedly, there was unprecedented fire damage.

The Malicious Damage Act Chap 11:06 creates an offence when a fire is maliciously set to any crop, hay grass, trees, wood or any other matter that is used for fuel.; at sections 19 and

20, when trees, saplings, shrubs or underwood valued at more $ 500 are damaged, the penalty is imprisonment for two years.

Table 11.1. (a) Areas burnt and (b) List of fires which occurred in 2010 in Trinidad, West Indies

AREA BURNT BY LAND USE IN TRINIDAD 1998 - 2008

Year	Natural Forest		Shrub/Sec. Forest		Teak Plantation		Pine Plantation		Savannah / Grasses		Other		Agricultural Lands		Grand Total	
	No. of Fires	Area of burnt (HA)	No. of Fires	Area of burnt (HA)	No. of Fires	Area of burnt (HA)	No. of Fires	Area of burnt (HA)	No. of Fires	Area of burnt (HA)	No. of Fires	Area of burnt (HA)	No. of Fires	Area of burnt (HA)	No. of Fires	Area of burnt (HA)
1998	21	248	96	845	100	5,467	51	584	327	1,732	49	182	118	1,225	**762**	**10,283.0**
1999	6	9	16	28	22	562	12	79	96	280	6	20	14	15	**172**	**993.0**
2000	-	-	8	18	19	717	8	23	39	136	3	9	15	24	**92**	**927.0**
2001	18	126	82	453	74	2,246	34	223	177	850	14	121	64	238	**463**	**4,257.0**
2002	-	-	7	15	4	9	2	10	41	85	4	10	4	5	**62**	**134.0**
2003	16	1,267	45	452	45	1,192	45	760	144	750	27	165	25	137	**347**	**4,723.0**
2004	-	-	13	175	19	968	3	18	76	276	17	32	8	16	**136**	**1,485.0**
2005	3	11	26	55	35	936	13	22	149	410	15	177	29	86	**270**	**1,696.0**
2006	1	15	21	55	16	606	13	153	106	338	53	79	0	0	**210**	**1,245.8**
2007	12	147	53	229	57	1,338	18	219	227	1,316	24	115	61	203	**452**	**3,567.0**
2008	2	8	23	43	18	402	16	135	83	704	10	28	74	216	**226**	**1,536.0**

(a)

Listed below is a summary of fires as at April 16, 2010

Conservancy	No. of Fires	Area (ha)		Total Area (ha
		State	Private	
North East	68	44.8	176.15	220.95
North Central	221	392.45	472.75	865.2
North West	98			986
South East	83	198.8	153.02	351.82
South Central	43	1755.6	265	2020.6
South West	29	1774.8		1774.8
Total	542	(4166.45)	(1066.92)	6219.37

(b)

Source: Forestry Division.

The Forest Act Chap 66:01 protects a wide variety of trees. The minister has the power to appoint forest officers and to confer with the Minister of National Security to use members of the Coast Guard as Forest Officers authorising them to prevent the removal of forest produce along the coast of Trinidad and Tobago. At section 7, any person wanting to fell a tree that is listed in the second schedule to the Act as well as any person who fells any tree on a slope of over 30 degrees within a hectare of land without a felling permit is liable on conviction to a fine of $20,000. No person can remove timber even from his own land without a permit.

There is a fine of twenty thousand dollars if the act was committed in a Forest Reserve or State lands.

The Sawmills Act at Chapter 66:02 goes a step further and requires that every person operating a furniture shop where a furniture shop, is defined as a place where wooden furniture is manufactured for sale, must have a permit issued by the Minister.

3.7. The State Lands Act

The State Lands Act Chapter 57:01 makes it an offence for any person to dig or remove material from State lands in Trinidad and Tobago without a license and where the material removed is asphalt, the fine is $250,000 and a term of imprisonment for two years. Where the material is not asphalt the penalty is a term of imprisonment for six months and a fine of $120,000.00. Where the Commissioner of State Lands has reasonable cause to suspect that material has been removed from State lands, he may seize the material and arrest the offender. The onus is on the person to prove to the Magistrate that that the material seized was not dug, won, or removed from State Lands without a licence in the prescribed form; and that he was not in fact engaged or concerned in the digging or removing of material from State Lands, or that he had a license in the prescribed form,

3.8. The Litter Act

This piece of legislation deals with the disposal of litter and waste particularly in public places and disposal of litter from motor vehicles. The Act empowers every policeman, forest warden, municipal officer, public health inspector, and harbour master to act as litter prevention wardens.

Where any warden finds a person depositing litter in a public place or has good cause to believe that a person has deposited any litter in any such place, the warden may require him to place the litter in the nearest receptacle or, if by reason of the size or quantity of the litter or for any other reason it is not practicable for the litter to be placed in a receptacle, to remove the litter to the nearest approved site.

An officer under this act may enter premises within reasonable hours for the removal of litter where he believes litter is left or caused to be kept in or on any premises in such circumstances as to cause, contribute to or tend to the defacement thereof by such litter. The local authority may give notice (whether orally or in writing) to the person so leaving or causing the same to be left or to the owner or occupier of such premises requiring him forthwith in the case of dead animals, carrion or other litter considered by the local authority to be dangerous to health or life, and in any other case within such time as may be limited by the notice (not being less than three days) to remove the litter so as to restore the premises to a condition satisfactory to the local authority.

Court may, in addition to imposing a penalty, order the offender, under the supervision and to the satisfaction of a person nominated by the Court, to clean up and remove such litter deposited in such public place within such period and to such place as may be specified in the order.

3.9. The Environmental Management Authority Act

The preamble to the EMA Act Chap 35:05 states the government's commitment to developing a national strategy for sustainable development, being the balance of economic growth with environmentally sound practices, in order to enhance the quality of life and meet the needs of present and future generations:

Management and conservation of the environment and the impact of environmental conditions on human health constitute is a shared responsibility and requires co-operation and co-ordination of public and private sector activities:

While several public authorities and other institutions have been performing various environmental functions and services under existing laws, there is need for a co- ordinated approach to ensure the application of those laws is consistent with the Government's commitment.

The powers and functions of the authority are, inter alia to

- promote and encourage among all persons a better understanding and appreciation of the environment ;
- encourage the integration of environmental concerns into private and public decisions;
- ensure the establishment of an integrated environmental management system in which the Authority, in consultation with other persons, determines priorities and facilitates co- ordination among governmental entities to effectively harmonise activities designed to protect, enhance and conserve the environment ;
- develop and effectively implement written laws, policies and programmes
- enhance the legal, regulatory and institutional framework for environmental management

The functions and powers of the authority are to make recommendations for a National Environmental Policy; to develop and implement policies and programmes for the effective management and wise use of the environment, consistent with the objects of this Act; to co-ordinate environmental management functions performed by persons in Trinidad and Tobago ; to make recommendations for the rationalisation of all governmental entities performing environmental functions

4. Compliance

Individuals and corporations wishing to undertake activities that affect the landscape are required to obtain a Certificate of Environmental Clearance *and to* comply with the conditions and mitigation measures in any such certificate ; Where the Authority believes that a person is in violation of an environmental requirement, the Authority serves a written notice of violation which shall include a request that the person make such modifications to the activity within a specified time, as may be required to allow the continuation of the activity; or an invitation to the person to make representations to the Authority concerning the matters specified in the Notice.

Whenever the Authority believes that any person has violated any environmental requirement, or is engaged in any activity which is likely to result in a violation the Authority has powers to go even further.

These powers include:

- seeking a restraining order to prohibit the continued violation or prevent the activity which will likely lead to a violation;
- seeking an order for the closure of any facility or a prohibition against the continued operation of the developer

The Act further establishes the environmental commission which has powers to enforce orders and judgments, and the power to impose fines and punishment as the High Court of Justice. It also has jurisdiction to hear and determine appeals from decisions or actions of the Authority as specifically authorised under this Act.

4.1. Certificate of Environmental Clearance

The Certificate of Environmental Clearance (Designated Activities) Order 2001 outlines the activities that require a Certificate of Environmental Clearance (CEC). Among the 44 activities listed are those that are closely linked to flooding : the Establishment of waste water or sewage treatment facilities,

The establishment of land drainage and irrigation systems –

- The establishment, modification, or expansion (inclusive of associated works)of a land drainage or irrigation scheme for a parcel of land or more than 1 hectare during a two-year period:
- The establishment of a flood control system or a water supply impoundment for a parcel of land of more than 1 hectare during a two-year period.
- The realignment or modification of drainage or river systems. Establishment of water distribution systems, Establishment of surface impoundments, dams or reservoirs for storage of water, catchment, abstraction or treatment of potable/process water

Clearing, excavation, grading and land filling –

- The clearing, excavation, grading or land filling of an area of more than 2 hectares during a two-year period
- The clearing of more than one-half a hectare of a forested area during a two-year period
- The clearing, excavation, grading or land filling of any area with a gradient of 1:4 or more; Coastal or offshore construction or modification and dredging activities

In 2007 Activity 23 which reads " Establishment of a facility for non-metallic mining and processing- The establishment, modification, expansion, decommissioning or abandonment (inclusive of associated works) of a facility for the mining, processing or storage of clay, andesite, porcellanite, limestone, oil sand, sand(s), gravel or other non-metallic minerals" was amended to state "Establishment of a facility for non-metallic mining and processing- The establishment, modification, expansion, decommissioning or abandonment (inclusive of associated works) of a facility for the mining, processing or storage of clay, andesite, porcellanite, limestone, oil sand, sand(s), gravel or other non-metallic minerals of areas of one hundred and fifty acres or more".

The Minerals Act 2000 was passed by both Houses of Parliament but has not been assented to by the President.

5. Discussion

The laws of Trinidad and Tobago are adequate to deal with all the factors listed by the EMA as causes of flooding. The laws include a range of provisions to protect the environment and minimise the risks associated with bringing flood waters into communities and homes. For instance, controlling fires will leave the vegetation intact, reduce soil exposure, prevent high run-off, soil erosion, and slippage where the lands are on a slope. The Agriculture Fires Act, the Forest Act, the Malicious Damage Act and the Sawmills Act all contain provisions which should control the flow of water.

Despite these provisions reports from the Conservator show an unusually large number of fires for the year 2010. When we look at the rainfall pattern over the last ten years and the number of fires for the same period there appears to be overwhelming evidence that the scales are heavily tipped against a safe environment. The figures for 2010 suggest that the high incidence of fires may be linked to the lengthy period of drought. When the rains eventually came there was widespread flooding accompanied by panic and chaos for what was described as the worst floods ever. The bottom line is that instead of a reduction in flooding the situation has worsened.

Despite high fines detailed in the several pieces of legislation, no person has been charged under any of the Acts for starting a fire without permission. The Agriculture Fire Act allows for the appointment of a County Fire Officer who in turn may appoint fire guardians. The Act also allows the fire guardians to take preventive action in instances where conditions that will endanger others exist. He has power to command others to take remedial action. The Act then provides for persons to enforce the law. Similarly, the Forest Act allows for the appointment of Forest Officers to carry out duties under the Act and to prosecute forest offences which include, kindling outside of the fire season as well as felling trees on State Lands. Despite the available legislation there is difficulty in proving a case unless the perpetrator is caught *in flagrante delicto*, i.e., red-handed. In the State Lands Act, a person can be charged based on suspicion. The onus then is on the suspect to prove that he did not carry or remove the material. That approach should perhaps be adopted in these cases.

The Town and Country Planning Division (TPCD) is the main body responsible for determining land use and development. The Division is responsible for granting permission for variations in land use, buildings houses and numerous other activities related to development. The Act gives the Minister the final say. Even though there is a Town Planning Panel which advises the Minister, there is no duty for him to take advice from the Panel. Interestingly, the Panel is made up of 3-5 persons who are themselves appointed by the

Minister. These persons do not have to meet any particular qualification requirement relating to the environment or planning.

There are unauthorised housing settlements on the hillsides of Diego Martin in the North West peninsula and in many parts of the Northern Range [19] all resulting in increased flooding. The TPCD has identified shortage of staff as one of its major problems. While legislation gives the TPCD the right to demolish structures, the TPCD takes the violators to court rather than demolishing the structures [20]. This approach has failed to prevent unauthorised construction. There is clearly a weakness in the monitoring system that allows these settlements to go up unchecked, meaning that the law in itself is not a deterrent.

The Town and Country Planning Division has the power to delegate responsibilities and must do so in such a way that local authorities can question the basis on which development is taking place. Local authorities must not be left to assume that permission has been granted by Town and Country.

Where the Minister reverses a decision of the municipal authority, the Act should require that he offers some reason to the authority, for these are the persons who know and understand the layout of the community and might be better placed to make decisions relating to the community. Decisions such as these should be made by a properly constituted committee and not be based on the decision of a single person. The decision of the Minister is final.

The State Lands Act gives the Commissioner of State Lands detailed powers to deal with squatting and encroachment. He has the power to institute prosecutions through the Director of Public Prosecutions in order to maintain control over state lands.

The municipal corporations play an important role in reducing risk of flooding through their responsibility to keep watercourses clear. A high percentage of flooding incidents are related to large amounts of water flowing into blocked drains and rivers and the failure to prevent dumping in watercourses.

A number of activities that create risk of flooding require a Certificate of Environmental Clearance (CEC) from the Environmental Management Agency (EMA). Examples are: timber cultivation and extraction of logs, clearing, excavation, grading and land-filling; waterproofing, caulking, paving; land reclamation, coastal or offshore construction or modification and dredging activities; establishment of infrastructure for air, land or marine transport ; establishment of surface impoundments, dams or reservoirs for storage of water and establishment of land drainage and irrigation schemes.

What actually takes place is that when an application is made for a land development, the Town and Country Planning Division grants permission and the applicant is then required to apply to the EMA for a CEC. In addition, permission must be sought from the Water and Sewerage Authority, the Ministry of Works, the Fire Services and the Trinidad and Tobago Electricity Commission. There appears to be a connection between the roles of the various agencies which leaves room for violation.

The EMA has its own environmental police who monitor activities that come under its purview. It seems however that investigations take place largely after complaints are made to the authority by affected parties. The EMA then investigates to determine whether there has actually been a violation and then sends out a notice of violation. This system puts the onus of being vigilant on community groups and activists with the result that many developments have reached an advanced stage before the EMA becomes aware of a violation.

The table [21] below shows some of the data collected from the EMA on notices of violations. The information reveals that the offenders come from a broad spectrum of medium sized businesses, large developers and even state companies. It also indicates that the environmental lobby group has a role to play in ensuring best practices by developers.

In one notice of violation sent to a private developer of housing for the Housing Development Corporation, the EMA issued a stop order for violation of the "environmental requirement" specified in the CEC. The notice of violation specifically identified a number of issues related to flooding, among them the condition that "Flood control measures shall be used to prevent the occurrence of flooding due to increased run-off from the project site during the Site Preparation, Construction and Occupancy Phases".

In another notice submitted to a state company in the oil and gas sector, the company failed to submit the schedule of works for site preparation activities, an oil well abandonment programme and realignment works on the Vessigny River,

One private developer failed completely to apply for a Certificate of Clearance for land reclamation works, where the "water's edge lacked protective barriers...and material was observed falling into the sea".

Environmental inspectors only visited this site after the EMA received complaints. The land reclamation works had been ongoing for two weeks. The fines are no deterrent when one considers the amount of money that is made by these developers. In making a determination as to the quantum of a fine, the magnitude of the effects to the environmental over time must be considered.

While there are many pieces of legislation in place for dealing with factors that affect flooding, there is little enforcement of the laws.

Because flooding is recognised as a massive problem political will is necessary for dealing with the problem. There must be comprehensive legislation relating to the causes of flooding. In addition, one body which coordinates the work of all the agencies involved is necessary so that work carried out in relation to flooding prevention can have best effect.

A good example of a comprehensive legislation can be found in Scotland In 1961 Scotland passed a Flood prevention Act. Now, the Flood Risk Management (Scotland) Act 2009 which was enacted on June 16, 2009 repeals the former Act.

Specific measures within the Flood Risk Management (Scotland) Act 2009 include:

- A framework for coordination and cooperation between all organisations involved in flood risk management
- Assessment of flood risk and preparation of flood risk management plans
- New responsibilities for SEPA, Scottish Water and local authorities in relation to flood risk management
- A revised, streamlined process for flood protection schemes
- New methods to enable stakeholders and the public to contribute to managing flood risk , and;
- A single enforcement authority for the safe operation of Scotland's reservoirs

Table 11.2. Extracts of Notices of Violation issued by EMA, Republic of Trinidad & Tobago, in 2008 and 2009

Consent Agreement date	Administrative Order date	NV date	Compensation (TT$)	Type of violation	Type of activity
2010-06-16		2009-11-10	20 000.00	Construction without CEC;	Pig farm
2010-05-20		2009-01-16	91 666.67	Breaches of CEC	Maqueripe beach
2010-03-18		2009-08-03	4 833.34	Land development without CEC; cease; protect land	Land development
2010-03-01	2009-07-28	2009-03-23	6 500.00	Construction & operation without CEC; cease	Factory
	2009-10-13	2009-05-07	1 031 333.22	Land development without CEC; cease; protect/ restore land	Land development
	2009-09-17	2009-07-23	162 500.00	Land development without CEC; cease; protect/restore land	Quarry road
	2009-08-25	2009-05-18	113 333.36	Construction completed without CEC; compensation	Hotel
no date		2009-04-15	1 372 833.31+	Breaches; cease breaches	Land development
	2009-03-25	2009-02-04	111 833.32	Land development without CEC; cease; protect/ restore land	Land development
	2008-09-11	2008-06-24	3 083.33	Operating without CEC; cease	Auto workshop
	2008-05-06	2008-03-14	341 333.33++	Construction without CEC; cease; Env	Water bottling plant
	2010-01-14			Breaches	Power plant
	2009-12-16		Operating w'out	CEC	Wood workshop

The Act creates a framework in which the organisations involved in flood risk management coordinate actions to deliver sustainable approaches to managing the causes and consequences of flooding. Central to this work will be the preparation of new assessments and plans to coordinate flood management actions.

To support the implementation of the Act and the development of accompanying legislation and policies, the Scottish Government has established a Scottish Advisory and Implementation Forum on Flooding (SAIFF). SAIFF comprises representatives from the government, government agencies, local authorities and other stakeholders. In addition to supporting the work of the Scottish Government, SAIFF will provide technical support to all parties involved in the implementation of the Act and the EC Floods Directive.

Conclusion

Comprehensive legislation – such as that found in the Flood Risk Management (Scotland) Act 2009 - relating to the causes of flooding is necessary, and one body should coordinate the work of all the agencies involved, so that work carried out in relation to flooding prevention can have the best effect.

References

[1] http://www.worldwatch.org/node/6267, Ben Block September 25th, 2009
[2] http://www.nature.com/ngeo/journal/v2/n10/abs/ngeo629.html, *Nature Geoscience*, 2, 681 - 686 (2009) Sinking deltas due to human activities James P. M. Syvitski1, Albert J. Kettner1, Irina Overeem1, Eric W. H. Hutton1, Mark T. Hannon1, G. Robert Brakenridge2, John Day3, Charles Vörösmarty4, Yoshiki Saito5, Liviu Giosan6 & Robert J. Nicholls7
[3] http://www.bbc.co.uk/news/world-south-asia-10834414
[4] http://articles.latimes.com/2010/jul/21/world/la-fg-china-floods -20100722
[5] http://www.caribbeannetnews.com/2005/01/17/flooding.shtml
[6] a b Table A.2, Database documentation, Latin America and the Caribbean (LAC) Population Database, version 3, International Center for Tropical Agriculture et al., 2005. Accessed on line February 20, 2008
[7] Caribbean Environment Outlook, United Nations Environment Programme, Copyright 1999 Page 17
[8] The New Oxford Dictionary of English, Oxford University Press, 1998
[9] Vol. 12 No. 1 EMA News
[10] Source discussions at Coastal Defense The Guianas, Paramaribo, Suriname, September 2010
[11] Clerk and Lindsell on torts (19th Ed) at para 20-06 Sweet and Maxwell
[12] Sedleigh-Denfield v O'Callaghan. [1940] AC 880at 903 per Lord Wright
[13] [1917]AC 556 at 572
[14] [2004] EWCA Civ 892;[2005] Env LR 6
[15] (1828) 8 B & C 355.
[16] (1886) LR 3 HL 330
[17] Tennent v Earl of Glasgow. (1864) 2 M Hl 22
[18] (1876) 2 Ex. D 1
[19] Report of the committee on Northern Range Hillside Development Policy, October 1988
[20] Newsday August 29th, 2010
[21] Information released by EMA after Freedom of Information Act application; summary compiled by P.Vine 2010-7-23

In: Flooding and Climate Change
Editors: D. D. Chadee, J. M. Sutherland and J. B. Agard
ISBN: 978-1-62948-317-7

Chapter 12

Climate Change Scenarios, Future Projections for Precipitation and Implications for Flooding in the Caribbean

Kishan Kumarsingh[1*] and Kenneth Kerr[2]
[1]Ministry of the Environment and Water Resources,
Port of Spain, Trinidad, West Indies
[2]Trinidad and Tobago Meteorological Services,
Trinidad, West Indies

Abstract

The scenarios used in previous climate change modeling could not incorporate mitigation or adaptation policies or strategies that may have been developed as a result of political or legislative intervention. However, the RCP approach can be ostensibly used to assess policy interventions as the new scenarios can respond to such policy and legislative decisions. The RCPs were developed as a result of an increasing interest in scenarios that can examine the impact of different climate policies [7], in addition to the no-climate-policy scenarios (e.g., as in SRES). Such scenarios would allow a cost - benefit analysis of long-term climate goals. It is to be noted that the development of the RCPs is not part of the IPCC process but that of the research community. The particular meteorological conditions leading up to a recent extreme (December 2013) are described, and presented as an example of the type of extreme event that is likely to increase in the region.

* Corresponding author: Head, Multilateral Environmental Agreements, Ministry of the Environment and Water Resources, Trinidad and Tobago.

INTRODUCTION

This chapter is a more detailed description of the scenarios used for projecting precipitation associated with projected climate change. Accordingly, a discussion of future precipitation is presented and the possible implications for flooding. Additionally a case study is presented of severe flooding episodes in the eastern Caribbean in late 2013 and how this may or may not reflect the projections made by climate models under various scenarios.

Scenarios and Models Used in Projecting Future Climate and Impacts

The Inter Governmental Panel on Climate Change (IPCC) has published its 5th Assessment report on the Science of Climate Change [1], which collates and analyses all published, peer reviewed scientific publications on climate change science. The IPCC also provides projections of climate change and associated changes in climate parameters such as, *inter alia*, temperature, precipitation and sea level rise. In order to do so, various scenarios of a future world are developed, taking into account specific assumptions, and which are all modeled using climate modeling software. It would therefore be useful to understand the basic approaches to projecting future climate change.

Scenarios

What Is a Scenario and Why is it Necessary to Develop Scenarios?

The IPCC has defined a scenario as "a coherent, internally consistent and plausible description of a possible future state of the world. It is not a forecast; rather, each scenario is one alternative image of how the future can unfold. A projection may serve as the raw material for a scenario, but scenarios often require additional information (e.g., about baseline conditions). A set of scenarios is often adopted to reflect, as well as possible, the range of uncertainty in projections". [2].

It is important to understand that a scenario is not a prediction of the future state of the world but are developed using assumptions that can be based on existing trends, such as greenhouse gas emissions, population growth and demographics and socio-economic trends. In the past, the IPCC developed the IS92 family of scenarios that were largely based on long term greenhouse gas emissions, technology development, and global economic and population growth. The IS92 scenarios were succeeded by the SRES (Special Report on Emission Scenarios) of the IPCC [3], which were constructed to explore future developments in the global environment with special reference to the production of greenhouse gases and aerosol precursor emissions, and were used in the IPCC's Third and Fourth Assessment Reports. The SRES scenarios used socio-economic 'storylines'[1] from which emission trajectories and climate impacts are projected.

At an initial level therefore, scenarios can be considered to be the common platform for scientists and policy makers to develop and understand the implications of future climate change.

---

[1] The IPCC (2) defines a storyline as "a narrative description of a scenario (or a family of scenarios), highlighting the main scenario characteristics and dynamics, and the relationships between key driving forces".

In 2007, the IPCC developed a new set of scenarios referred to as the Representative Concentration Pathways or RCPs, which are used in the Fifth Assessment Report in preference to the SRES.

The IPCC describes these scenarios as follows [4]:

> "In climate change research, scenarios describe plausible trajectories of different aspects of the future that are constructed to investigate the potential consequences of anthropogenic climate change. Scenarios represent many of the major driving forces - including processes, impacts (physical, ecological, and socioeconomic), and potential responses that are important for informing climate change policy. They are used to hand off information from one area of research to another (e.g., from research on energy systems and greenhouse gas emissions to climate modeling). They are also used to explore the implications of climate change for decision making (e.g., exploring whether plans to develop water management infrastructure are robust to a range of uncertain future climate conditions). The goal of working with scenarios is not to predict the future but to better understand uncertainties and alternative futures, in order to consider how robust different decisions or options may be under a wide range of possible futures".

The IPCC further states [4]:

> "Scenarios of different rates and magnitudes of climate change provide a basis for assessing the risk of crossing identifiable thresholds in both physical change and impacts on biological and human systems".

This description underscores the utility value of developing appropriate scenarios in the context of addressing climate change, and the ability to conduct cost -benefit analysis of alternative strategies for effective policy development and planning, and instituting appropriate management structures with an understanding of the associated uncertainties. *This aspect of the use of scenarios is the most critical.*

International research on climate change explores different aspects including, *inter alia,* implications for biological, chemical, physical and socio-economic systems, and their interaction at primary, secondary and tertiary levels (Figure 12.1, [adapted from IPCC 2007 [4]]. In order for the results of various impacts to be comparable, studies would have to be based on some set of common basic parameters and assumptions, including baselines. These can be considered to be the rudiments of scenario development and climate change modeling.

Representative Concentration Pathways or RCPs Compared to SRES Scenarios

This section is not intended to provide a detailed description of the development of the RCPs but to give an overview. The RCP scenarios differ from the previous IPCC scenarios (IS92 and SRES families). In earlier scenario development, future greenhouse gas and aerosol emissions were derived from socio-economic scenarios, and which were then used to assess the impacts of those emissions on the climate system and the implications of the climate change s, as well as changes in socio-economic and environmental futures, on human and natural systems. This entire process has been shown to take about 10 years [5].

The approach to RCP development is the other way around. Instead of using socio-economic scenarios as the starting point to derive alternative greenhouse gas emissions, the

RCP scenarios use alternative futures in global greenhouse gas and aerosol concentrations as their starting point. The RCPs not only describe future greenhouse gas and aerosol concentrations, but also incorporates a *trajectory* to these concentrations by the year 2100. The usefulness of this approach is that each trajectory is constructed from a synthesis of the published literature - a sort of 'baseline" development, against which various combinations or permutations of social, technical and economic circumstances, (which are referred to as 'narratives ' and the "storyline" equivalent of the SRES scenarios) can be tested. The purpose of developing RCPs is to shorten the time to develop and apply new scenarios, but perhaps more importantly, to ensure better integration between socio-economic driving forces, changes in the climate system, and the vulnerability of natural and human systems. Accordingly, it is intended that the RCPs can be used in parallel [5]:

- by Earth System Models (ESMs) to explore future changes in physical and biogeochemical responses to changing atmospheric composition and radiative forcing, and
- by Integrated Assessment Models (IAMs) to explore alternative socio-economic conditions that would result in such future atmospheric composition changes.

and to generate climate model that will be used to construct new climate scenarios for application in Impacts, Adaptation, and Vulnerability (IAV) and IAM studies. The new process therefore intended to be used by climate modelers and other scientists in the IAM and IAV communities to afford comparison of their results, and facilitate the use of new climate modeling results in conjunction with IAV research [6].

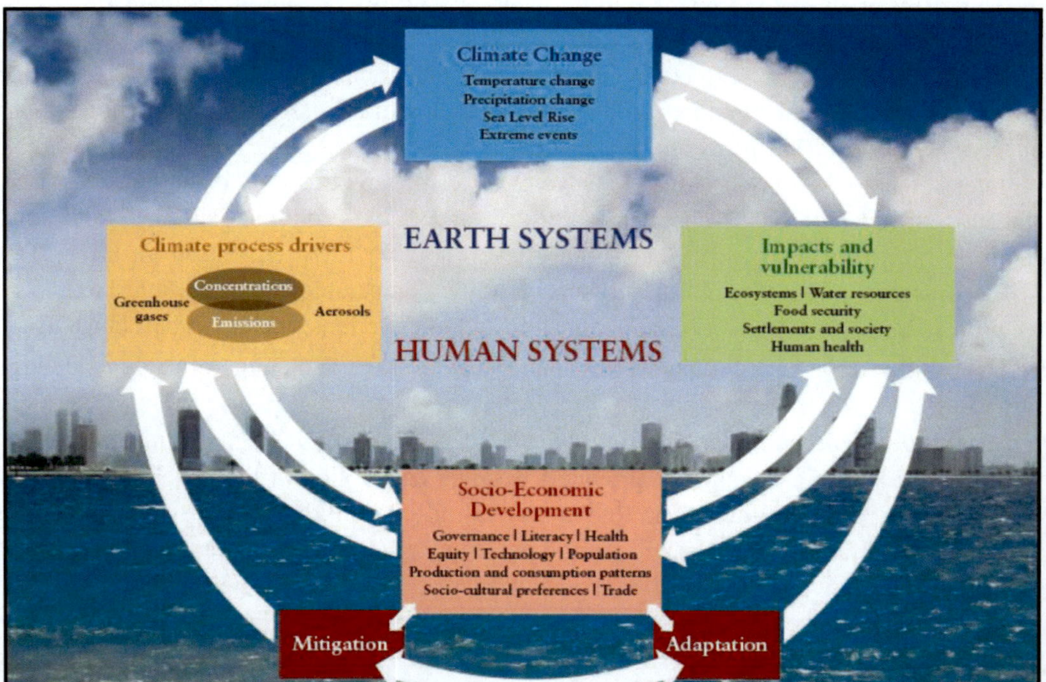

Figure 12.1. Inter-relationship between human and Earth systems as illustrated by the IPCC [4].

Previously, the scenarios could not incorporate mitigation or adaptation policies or strategies that may have been as a result of political or legislative intervention. However, the RCP approach and be ostensibly used to assess policy interventions as the new scenarios can respond to such policy and legislative decisions. The RCPs were developed as a result of an increasing interest in scenarios that can examine the impact of different climate policies [7], in addition to the no-climate-policy scenarios (e.g., as in SRES). Such scenarios would allow a cost -benefit analysis of long-term climate goals. It is to be noted that the development of the RCPs is not part of the IPCC process but developed by the research community. The scenarios were used in the IPCC Fifth Assessment Report.

Projections of the IPCC's Fifth Assessment Report - Precipitation

Projected changes in the climate are classified in the Fifth Assessment Report by the IPCC as probabilities.[2] The figures 12.2 and 12.3 are taken from the IPCC reports and serves only to illustrate the general projections. No attempt is made to describe details of the models used or to provide an evaluation of the models. The reader is referred to relevant publications for a deeper understanding of the information required and the process by which assessments are conducted.

Precipitation

Change in average global precipitation (1986−2005 to 2081−2100) is illustrated in Figure 12.2 [8].

Projections for increase in the frequency, intensity, and/or amount of heavy precipitation is very likely over most of the mid-latitude land masses and over wet tropical regions. The projections for the Central American and Caribbean region is illustrated in Figure 12.3 [9].

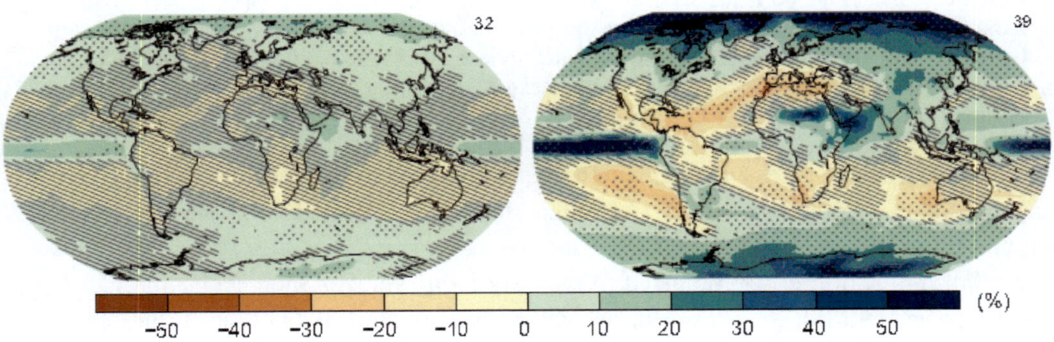

Figure 12.2. "Change in average precipitation (1986−2005 to 2081−2100)" (taken from IPCC 2013).

---

[2] Virtually certain: 99–100% probability
Very likely: 90–100% probability
Likely: 66–100% probability
About as likely as not: 33–66% probability
Unlikely: 0–33% probability
Very unlikely: 0–10% probability
Exceptionally unlikely: 0–1% probability
Additional terms (extremely likely: 95–100% probability, more likely than not: >50–100% probability, and extremely unlikely: 0–5% probability) may also be used when appropriate.

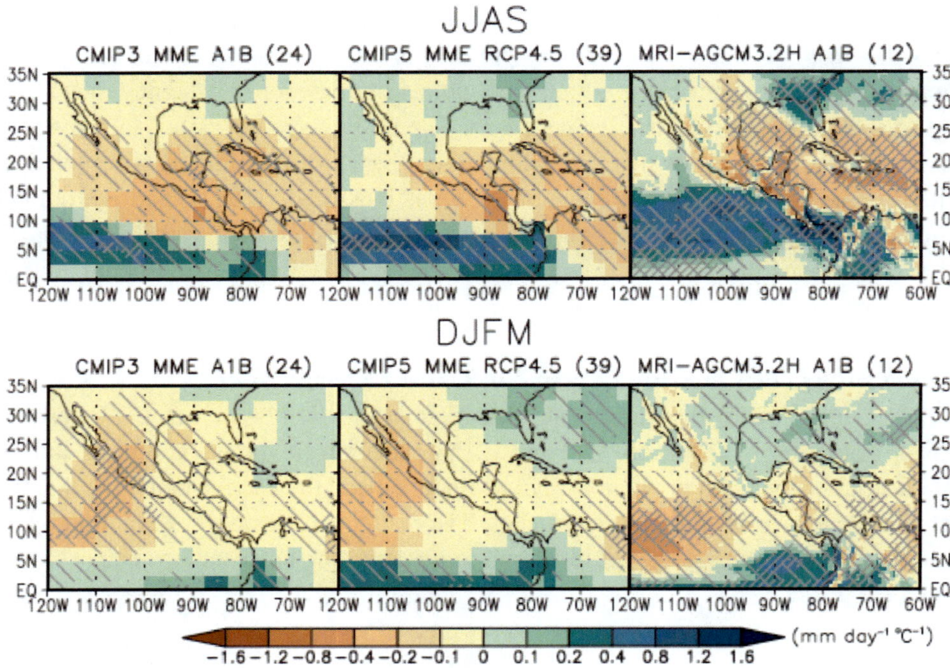

Figure12.3. "Maps of precipitation changes for Central America and Caribbean in 2080–2099 with respect to 1986–2005 in June to September (above) and December to March (below) in the SRES A1B scenario with 24 CMIP3 models (left), and in the RCP4.5 scenario with 39 CMIP5 models (middle). Right figures are the precipitation changes in 2075–2099 with respect to 1979–2003 in the SRES A1B scenario with the 12 member 60 km mesh Meteorological Research Institute (MRI)-Atmospheric General Circulation Model 3.2 (AGCM3.2) multi-physics, multi-SST ensembles (Endo et al., 2012). Precipitation changes are normalized by the global annual mean surface air temperature changes in each scenario. Light hatching denotes where more than 66% of models (or members) have the same sign with the ensemble mean changes, while dense hatching denotes where more than 90% of models (or members) have the same sign with the ensemble mean changes" (taken from Christensen et al., 2013, IPCC Report 2013)[9]).

The projected changes in precipitation and extreme event can have implications for flooding in the Caribbean region. Although there are projected decreases in cumulative rainfall, the likelihood of increased intensity and extreme events can mean a greater amount of rainfall in a short period compared to the long term average. Such events, based only on climate considerations, may result in an increased likelihood of flooding in the regions. A case study in presented below on actual observed recent extreme weather events and how it reconciles with regional climate projections.

CASE STUDY

The Recent Extreme Rainfall Event in the Eastern Caribbean in Late 2013

Usually, much attention is given to the impact of hurricanes in the Eastern Caribbean region due to the destruction left by these systems. However, recently extreme heavy rainfall events that are not associated with hurricanes in the region are causing just as much destruction and damage. On December 24-25 2013, one of these extremely heavy rainfall events caused significant destruction in St Lucia, St Vincent and Dominica. This event can be considered an extreme event due to the excessively high rainfall amounts (6 inches in 24 hrs) and strong winds it generated and the high damage cost which resulted, including 15 deaths [3].

Generally, rainfall towards the end of the wet season in the eastern Caribbean is greatly influenced by pendulum-swinging mid-Atlantic mid to upper level low pressure and trough systems, with concomitant strong upper- level wind divergence which draws air upward, providing a lifting mechanism for the formation of cloud with large vertical extent which produces voluminous rainfall amounts. The meteorological conditions which existed 24 hours prior to and during the occurrence of the extreme event in late December were precursors for an extreme rainfall event. These conditions at the time were: a low pressure center throughout a deep tropospheric layer with concomitant trough axes just west of the islands ; upper level wind speed maxima and divergence above and to the east of the islands; and a collocated migrating late season surface to low level trough in the vicinity of the islands, with high equivalent potential temperatures and high moisture content converging and being transported by a strong low level wind jet into the area.

In the past, the region has experienced similar extreme rainfall events, for instance in Jamaica : January 3-4, 1998 a similar trough system produced heavy rainfall which resulted in 5 deaths and 9 million $US in damage[4]; Puerto Rico : January 5-6, 1992, a comparable surface trough resulted in 23 deaths and 88 million $US[5]; Trinidad and Tobago : October 9-10, 1981, a complex weather system produced 6 inches of rainfall in 24 hours with wide spread flooding and high damage cost and in August 11, 2012, another system resulted in high damage costs and 2 deaths [6]. The recent intense weather extreme therefore fits into a larger account of observed similar damaging intense weather events in the last three decades in the region. The latest event can be viewed as providing a snapshot of a larger statistical trend of more frequent and intense extreme precipitation events and/or amount of heavy rainfall likely observed over land areas [IPCC, 2013, [8]].

The event also fits into downscaled regional climate model outputs for Trinidad and Tobago which suggest that in a warming environment there is likely to be increases in intensity and frequency of rainfall related to thermal convection and moisture convergence which drives deep convective clouds development [Kerr, 2012, [10]], under an A2 SRES future scenario [11]. It is therefore reasonable to expect that under an IPCC A2 SRES[7] future scenario, extreme rainfall events could become more frequent in Trinidad and Tobago and the Southern Caribbean Region. It is important to state that the recent event cannot be conclusively attributed to climate change, since the evidence shows that similar extreme events have occurred in the region in the absence of the attribution of climate change. However, based on the knowledge and understanding of climate physics which suggests that

---

[3] Caribbean Disaster Emergency Management Agency : www.cdema.org

[4] Arlene Laing and Paul Saunders (2002) Flash Floods in Jamaica during January and March 1998; Environmental conditions and mitigation. American Meteorological Society Symposium on Precipitation Extremes.

[5] NOAA, National Weather Service, 1992: Puerto Rico Flash Floods January 5-6, 1992. National Disaster Survey Report, Silver Springs, Maryland, 32pp.

[6] Trinidad and Tobago Meteorological Service Event Report

[7] IPCC, 2007: Summary for Policymakers. In: Climate Change 2007: The Physical Science Basis. Contribution of Working Group I to the Fourth Assessment Report of the Intergovernmental Panel on Climate Change [Solomon, S., D. Qin, M. Manning, Z. Chen, M. Marquis, K.B. Averyt, M. Tignor and H.L. Miller (eds.)]. Cambridge University Press, Cambridge, United Kingdom and New York, NY, USA.

ongoing warming, if continued, will likely result in increased intensity and frequency of extreme weather events, allow the event described above to be used as an example of the type of extreme weather events that are likely to increase in the region under a warming future.

CONCLUSION

The knowledge of the meteorological conditions which led to the extreme weather event described above will enable forecasters to provide earlier warning in future; and, in the longer term and the wider context, continued climate modeling and use of scenarios such as RCPs will enable cost -effective adaptation and mitigation solutions.

REFERENCES

[1] Hartmann, D. L., Klein Tank, A. M. G., Rusticucci, M., Alexander, L. V., Brönnimann, S., Charabi, Y., Dentener, F. J., Dlugokencky, E. J., Easterling, D. R., Kaplan, A., Soden, B. J., Thorne, P. W., Wild, M. & Zhai, P. M. (2013). *Observations: Atmosphere and Surface*. In: Climate Change 2013: The Physical Science Basis. Contribution of Working Group I to the Fifth Assessment Report of the Intergovernmental Panel on Climate Change [Stocker, T.F., D. Qin, G.-K. Plattner, M. Tignor, S.K. Allen, J. Boschung, A. Nauels, Y. Xia, V. Bex and P.M. Midgley (eds.)].Cambridge University Press, Cambridge, United Kingdom and New York, NY, USA.

[2] http://www.ipcc-data.org/guidelines /pages/definitions.html

[3] IPCC. (2000). - Nebojsa Nakicenovic and Rob Swart (Eds.) Cambridge University Press, UK. 570

[4] Towards New Scenarios for Analysis of Emissions, Climate Change, Impacts, and Response Strategies. *IPCC Technical Summary, 2007* http://www.ipcc.ch/scoping _meeting_ar5/expert-meeting-report-scenarios.pdf

[5] http://sedac.ipcc-data.org/ddc/ar5_scenario_process/index.html

[6] http://www.globalchange.umd.edu/iamc_data/sma-ssp/02Scenario_Framework_ Changwon_110716.pdf

[7] Moss, R. H., Edmonds, J. A., Hibbard, K. A., Manning, M. R., Rose, S. K., van Vuuren, D. P., Carter, T. R., Emori, S., Kainuma, M., Kram, T. et al. (2010). The next generation of scenarios for climate change research and assessment. *Nature*, *463*, 747–756

[8] IPCC. (2013). Summary for Policymakers. In: Climate Change 2013: The Physical Science Basis. Contribution of Working Group I to the Fifth Assessment Report of the Intergovernmental Panel on Climate Change [Stocker, T.F., D. Qin, G.-K. Plattner, M. Tignor, S.K. Allen, J. Boschung, A. Nauels, Y. Xia, V. Bex and P.M. Midgley (eds.)]. Cambridge University Press, Cambridge, United Kingdom and New York, NY, USA.

[9] Christensen, J. H., Krishna Kumar, K., Aldrian, E., An, S.-I., Cavalcanti, I. F. A., de Castro, M., Dong, W., Goswami, P., Hall, A., Kanyanga, J. K., Kitoh, A., Kossin, J., Lau, N.-C., Renwick, J., Stephenson, D. B., Xie, S.-P. & Zhou, T. (2013). Climate Phenomena and their Relevance for Future Regional Climate Change. In: Climate

Change 2013: The Physical Science Basis. Contribution of Working Group I to the Fifth Assessment Report of the Intergovernmental Panel on Climate Change [Stocker, T.F., D. Qin, G.-K. Plattner, M. Tignor, S.K. Allen, J. Boschung, A. Nauels, Y. Xia, V. Bex and P.M. Midgley (eds.)]. Cambridge University Press, Cambridge, United Kingdom and New York, NY, USA.

[10] Kerr, K. (2012). Capturing Recent Climate Variations in the Extreme Southern Caribbean with Projections for the Near and Medium Terms: A Downscaling Approach for Trinidad and Tobago. MSc Thesis, University of Reading, 77pp.

[11] IPCC, (2007b). Summary for Policymakers. In Climate Change 2007: The Physical Science Basis. Contribution of Working Group I to the Fourth Assessment Report of the Intergovernmental Panel on Climate Change (IPCC AR4) S. Solomon et al. eds (Cambridge University Press).

INDEX

A

B

C

F

G

H

N

O

P

Q

R

S

T

U

V

W

Z